Hampton Roads Naval Aviation

A CENTENNIAL RETROSPECTIVE

The Virginian-Pilot

 • ISBN: 978-1-59725-342-0

Published by Pediment Publishing, a division of The Pediment Group, Inc. www.pediment.com Printed in Canada.

Foreword

In the 100 years since a barnstorming pilot made his first nail-biting takeoff from a cruiser in Willoughby Bay, Hampton Roads has had a front-row seat for the drama that is naval aviation.

We built the aircraft carriers. We trained the pilots. We fueled and maintained and fixed the aircraft – and scrapped them, too, when their flying days were done.

By some measures, the pace of change over the century seems, well, supersonic. Try to imagine what that first daring pilot, Eugene Ely, would think if he flew with the Blue Angels at the Oceana Air Show today, rolling and diving through the skies of Virginia Beach. That is, if he didn't pass out from the g-forces.

But some constants remain. Despite email and Skype, sailors at sea still anticipate the sound of the carrier onboard delivery plane bringing the mail. Anxious spouses and kids, flags in hand, still crowd the pier and the flight lines at Oceana and Chambers Field – though now it may be Mom rather than Dad who emerges from the cockpit or the brow. And the mystical combination of sky and sea still calls to those with a spirit of adventure and a desire to serve their country.

So come see where we've been, and where we're headed. Blimps and bombers, helicopters and Hellcats and Hornets – you'll see them all in these pages, and meet many of the pilots who flew them. Newspaper front pages will show how events across the world hit home here. And family photos shared by Pilot readers will add a unique, personal touch to the story.

This book would not have been possible without the contributions of a number of individuals and institutions. Special thanks go to the staffs of the Hampton Roads Naval Museum, the Norfolk Public Library Sargeant Memorial Collection, and the Oceana Naval Air Station archives.

At The Pilot, news researchers Jaakon Hays and Maureen Watts helped find the treasures within our yellowed clip files and reels of microfilm. Reporters Corinne Reilly, Bill Sizemore and Kate Wiltrout wrote introductions to the chapters. And the work of many talented Pilot photographers graces these pages.

Finally, we're deeply grateful to the enlisted sailors and officers, past and present, who have allowed our journalists to share their stories.

RADIO
03
SB2C-5
NAVY

Table of Contents

3-F-18
3-F-17
3 16
3-F-6
3-F-12
3 F 13

CHAPTER ONE

The Early Years

IN 1910, THE CONCEPT OF naval aviation seemed far-fetched, if not downright crazy.

It had been only seven years since the Wright brothers' fragile biplane rose 10 feet above the sand of North Carolina's Outer Banks. Aviation was still a curiosity, an entertaining diversion with a circus side-show flavor. Most of the Navy's top brass were disdainful of the idea that these newfangled flying machines had any place in the sea service.

What began to change their minds was an improbable feat by a taciturn Midwesterner with a mechanical bent and a fearless streak.

Iowa-born Eugene Ely was not a Navy man. He was a civilian, one of the early barnstorming "birdmen" who thrilled crowds with airborne exploits at public exhibitions. He was barely 24 when he flew a 700-pound Curtiss "pusher" plane off a jerry-rigged Navy cruiser in Hampton Roads.

After an agonizing few seconds when it appeared the tiny aircraft would be swallowed by the waves, Ely ascended into the misty air. Five minutes later, he set the plane down alongside the weathered beach houses of Willoughby Spit on the northern tip of Norfolk – the first ship-to-shore flight.

Just after the United States entered World War I in 1917, the Navy acquired a 474-acre tract in Norfolk, a stone's throw from Ely's landing site. One of the first things built on the new base was an airfield for seaplanes.

The first U.S. Navy aviator was a Virginian, Theodore "Spuds" Ellyson, who, the story goes, set his career course as a boy when he saw a Navy fleet enter the Hampton Roads harbor.

The first U.S. aircraft carrier, the converted collier Langley, was commissioned in 1922. Carriers soon became a common sight at the Norfolk base, which went on to host the largest concentration of naval forces in the world.

OPPOSITE: Navy F3B fighters from a squadron attached to the carrier USS Lexington, at Norfolk Naval Air Station, 1920s.

Courtesy Sargeant Memorial Collection, Norfolk Public Library

ABOVE: Eugene Ely's Curtiss Pusher aircraft is serviced after being hoisted onto a temporary platform on the deck of the USS Birmingham at Norfolk Naval Shipyard in Portsmouth, November 1910.
Courtesy Hampton Roads Naval Museum

TOP RIGHT: Ely went on to make the first landing on a ship, the USS Pennsylvania, off the San Francisco waterfront in January 1911. He died in a plane crash later that year. Courtesy Hampton Roads Naval Museum

RIGHT: Ely's plane is hoisted onto the USS Birmingham in preparation for the first flight, November 1910. Courtesy Hampton Roads Naval Museum

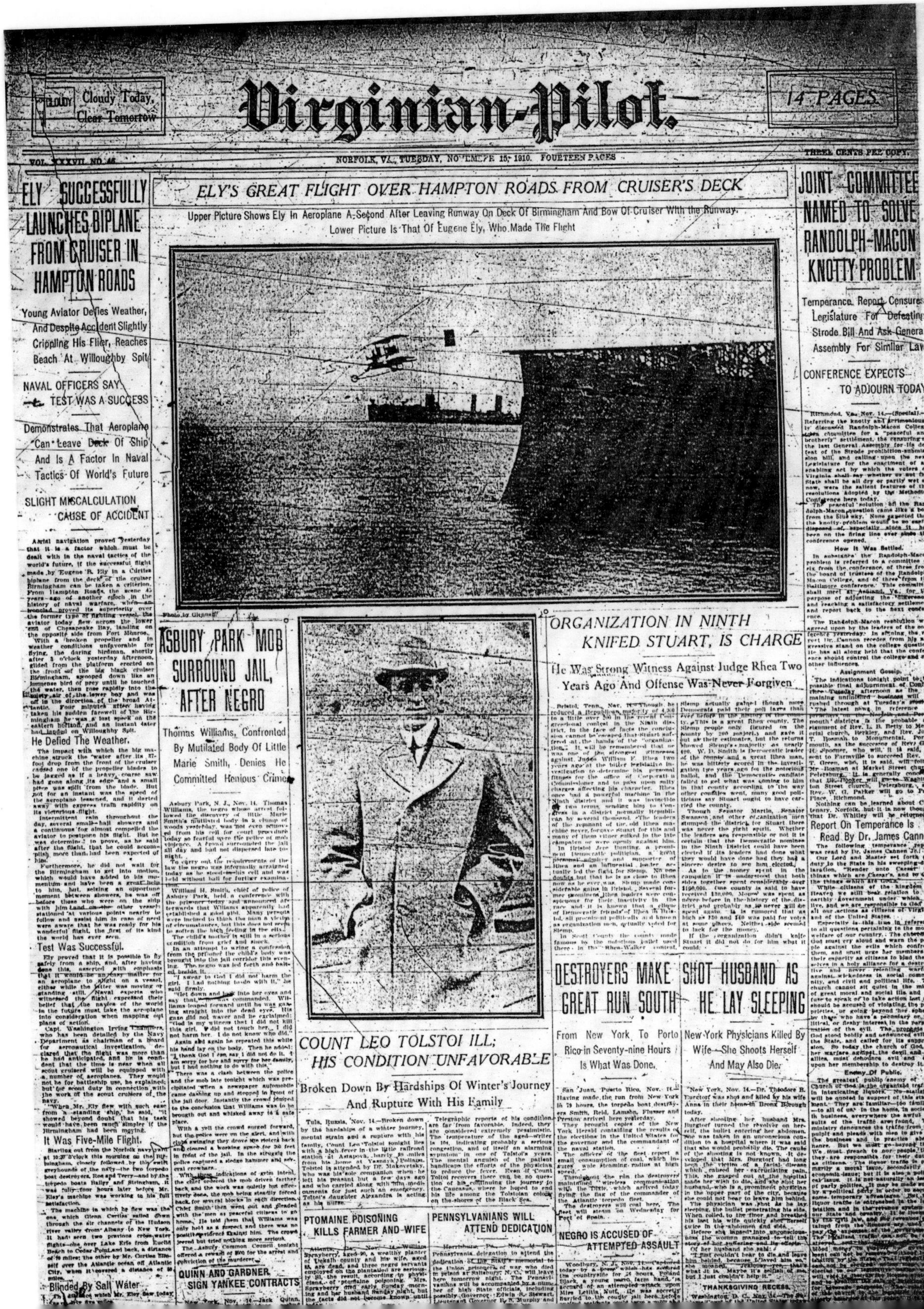

Virginian-Pilot.

Cloudy Today, Clear Tomorrow

14 PAGES

NORFOLK, VA., TUESDAY, NOVEMBER 15, 1910. FOURTEEN PAGES

THREE CENTS PER COPY.

ELY SUCCESSFULLY LAUNCHES BIPLANE FROM CRUISER IN HAMPTON ROADS

Young Aviator Defies Weather, And Despite Accident Slightly Crippling His Flier, Reaches Beach At Willoughby Spit

NAVAL OFFICERS SAY TEST WAS A SUCCESS

Demonstrates That Aeroplane Can Leave Deck Of Ship And Is A Factor In Naval Tactics Of World's Future

SLIGHT MISCALCULATION CAUSE OF ACCIDENT

He Defied The Weather.

Test Was Successful.

It Was Five-Mile Flight.

Blinded By Salt Water.

ELY'S GREAT FLIGHT OVER HAMPTON ROADS FROM CRUISER'S DECK

Upper Picture Shows Ely In Aeroplane A Second After Leaving Runway On Deck Of Birmingham And Bow Of Cruiser With the Runway. Lower Picture Is That Of Eugene Ely, Who Made The Flight

JOINT COMMITTEE NAMED TO SOLVE RANDOLPH-MACON KNOTTY PROBLEM

Temperance Report Censures Legislature For Defeating Strode Bill And Ask General Assembly For Similar Law

CONFERENCE EXPECTS TO ADJOURN TODAY

How It Was Settled.

Assignment Gossip.

Report On Temperance Is Read By Dr. James Cannon

Enemy Of Public.

ASBURY PARK MOB SURROUND JAIL, AFTER NEGRO

Thomas Williams, Confronted By Mutilated Body Of Little Marie Smith, Denies He Committed Heinous Crime

ORGANIZATION IN NINTH KNIFED STUART, IS CHARGE

He Was Strong Witness Against Judge Rhea Two Years Ago And Offense Was Never Forgiven

COUNT LEO TOLSTOI ILL; HIS CONDITION UNFAVORABLE

Broken Down By Hardships Of Winter's Journey And Rupture With His Family

DESTROYERS MAKE GREAT RUN SOUTH

From New York To Porto Rico in Seventy-nine Hours Is What Was Done.

SHOT HUSBAND AS HE LAY SLEEPING

New York Physicians Killed By Wife—She Shoots Herself And May Also Die.

PTOMAINE POISONING KILLS FARMER AND WIFE

PENNSYLVANIANS WILL ATTEND DEDICATION

NEGRO IS ACCUSED OF ATTEMPTED ASSAULT

QUINN AND GARDNER SIGN YANKEE CONTRACTS

THANKSGIVING RECESS.

ABOVE: Glenn Curtiss, manufacturer of the 'Triad' Amphibian, instructing Lt. Theodore G. "Spuds" Ellyson, Naval Aviator No. 1, at Hammondsport, N.Y., July 1911. The Triad became the Navy's first airplane. Courtesy Albert E. Grimstead

LEFT: Bad weather threatened to postpone Ely's flight, The Virginian-Pilot reported. But he was determined to forge ahead. Courtesy Virginian-Pilot archives

BELOW: Eugene Ely's plane plunges toward the water as it leaves the deck of the USS Birmingham, November 1910. Moments later, it skimmed the water before beginning to climb. Courtesy Hampton Roads Naval Museum

RIGHT: Pilot Eugene Ely is about to land on the deck of the USS Pennsylvania in his famous shore-to-ship flight, January 1911. Courtesy Albert E. Grimstead

BOTTOM RIGHT: Ely's landing was aided by the first arresting gear, a tailhook designed and built by aviator Hugh Armstrong Robinson. This photograph shows the Pennsylvania's flight deck with Ely's pusher loaded aboard. Courtesy The Amy Waters Yarsinske Collection

BELOW: Ely at the instant of his landing on the Pennsylvania. Courtesy Albert E. Grimstead

LEFT: Naval Aviator No. 1, Theodore Gordon "Spuds" Ellyson, a native of Richmond, Virginia, graduated in 1901 from the United States Naval Academy and started his military career aboard battleships and submarines. He was present at Newport News Shipbuilding when the submarine USS Seal was fitted out and served as its commanding officer just after its December 1910 commissioning. Then Ellyson was ordered to San Diego for flight instruction under Glenn Curtiss. Later in Ellyson's storied career, in January 1921, he was ordered to Hampton Roads for eight months as executive officer of the naval air station and after that headed back to Washington, D.C., to lead the plans division at the newly minted Bureau of Aeronautics. In June 1926, he returned to Norfolk, ordered to duty on the aircraft carrier USS Lexington (CV 2) for its fitting out and commissioning. Cmdr. Ellyson was killed on February 27, 1928, his forty-third birthday, when his aircraft crashed in the lower Chesapeake Bay on a night flight from Norfolk to Annapolis. He is shown here in 1911.
Courtesy Library of Congress Prints and Photographs Division Photo, Amy Waters Yarsinske Collection

ABOVE: Interior of Alfred C. Ward's home at 493 Breezy Point Crescent in Norfolk, circa 1915. Ward, seated, ran a general store and gas station at what became known as Ward's Corner in Norfolk. His home became the quarters for the commanding officer of Norfolk Naval Air Station. Courtesy Mary Lou Olson (Rockwell)

LEFT: Before the Navy had either planes or pilots, Capt. Washington I. Chambers arranged a series of tests in which Glenn Curtiss and Eugene Ely showed the world that aviation could go to sea. Chambers Field at Norfolk Naval Air Station was named in his honor. Courtesy Albert E. Grimstead

BELOW: Naval aviation machinist's mates at the first course of training at Aeronautic Station Hampton Roads, circa 1917. T.W. Hall was company commander. Courtesy Eric C. Beck

ABOVE: Hangar for lighter-than-air craft, Hampton Roads Naval Air Station, 1918. An airship is being inflated in the hangar. Courtesy Hampton Roads Naval Museum

BELOW: Fire from a hydrogen-filled airship that landed across high tension wires at Hampton Roads Naval Air Station, 1918. Courtesy Hampton Roads Naval Museum

ABOVE: An early view, circa 1918, of what would become part of Hampton Roads Naval Air Station (renamed Norfolk Naval Air Station in 1921). This area was originally developed for the Jamestown Exposition in 1907. Included in this view is Grand Basin, which faced Raleigh Square. The Japanese Bridge on the Government Pier is in the center of the photo. The Susan Constant Pier is in the foreground and Godspeed Pier in the distance. Courtesy Caroline Doonan

BELOW: Curtiss H-12 flying boats awaiting a call at Hampton Roads Naval Air Station, circa 1918. Only 19 H-12s were built. Courtesy Dick Hanna

LEFT: A pilot climbs a rope ladder to get aboard a kite balloon during a training exercise at Hampton Roads Naval Air Station, 1919. Courtesy Dick Hanna

BOTTOM LEFT: View of Hampton Roads Naval Operations Base, 1918. The airfield for seaplanes detached from the base and became Hampton Roads Naval Air Station in August of that year. Courtesy Sargeant Memorial Collection, Norfolk Public Library

OPPOSITE: Administrative heads of Hampton Roads Naval Air Station,1918. Courtesy Hampton Roads Naval Museum

BELOW: Non-rigid 'B' dirigible, Hampton Roads Naval Air Station, 1919. Courtesy Hampton Roads Naval Museum

ABOVE: The dirigible Roma exits the hangar at Norfolk Naval Air Station, for an ill-fated test flight, February 21, 1922.
Courtesy Hampton Roads Naval Museum

LEFT: An early seaplane at Hampton Roads Naval Air Station, 1920.
Courtesy Ocean View Station Museum

ABOVE: Postcard for the naval air station, circa 1920. Courtesy Dick Hanna

RIGHT: Rescue workers comb through the wreckage of the dirigible Roma at the naval air station, February 21, 1922. The fiery crash killed 34 of the 45 people onboard. It was after this accident that less-flammable helium gas replaced hydrogen in U.S. dirigibles.
Courtesy Hampton Roads Naval Museum

LEFT: Airmen use a pump to store helium gas from the Navy's C-7 airship at the Norfolk Naval Air Station, 1922. The airship was built in 1918 by Goodyear Tire & Rubber and Goodrich Rubber Company. It was 192 feet long and originally was filled with hydrogen. After the Roma crash, it was the first Navy blimp to be filled with helium. Courtesy Sargeant Memorial Collection, Norfolk Public Library

BOTTOM LEFT: One of the most unusual aircraft designs in the Navy inventory was the Gallaudet D-series, produced by the Gallaudet Engineering Company, Norwich, Conn.. In 1924, the company's founder, Edson Gallaudet, retired and sold his company's assets to Army Maj. Reuben Fleet who used them as the core around which he founded Consolidated Aircraft Corporation, which later became part of General Dynamics. Courtesy Hampton Roads Naval Museum

RIGHT: The rigid airship USS Shenandoah (ZR 1) conducts its first successful mooring with the fleet oiler USS Patoka (AO 9) in Norfolk, August 1924. Built at Newport News Shipbuilding, the Patoka was chosen in 1924 as a tender for the Shenandoah, Los Angeles (ZR 3) and Akron (ZRS 4). The ship was fitted with a mooring mast that rose about 125 feet above the water, with space to accommodate the airship's crews, supplies and equipment and with handling and stowage facilities for three seaplanes. In October 1925, Shenandoah broke up in flight during violent weather in Ohio; 14 of her air crew perished. Courtesy The Amy Waters Yarsinske Collection

BOTTOM RIGHT: The administration building at the naval air station, 1920s. Courtesy Dick Hanna

BELOW: Ralph B. Hanna Sr., feeding Whiskey, a bear captured as a cub by a pilot who crashed in the Great Dismal Swamp. The cub was blind and had been orphaned by the crash. It became the air station's mascot and was featured at Saturday "smokers," where it would wrestle with volunteers from the audience. If a sailor held an Almond Hershey candy bar in his teeth, Whiskey would try and get it. No one was afraid of Whiskey but no one chose to argue with him. Courtesy Dick Hanna

ABOVE: The USS Langley, the Navy's first aircraft carrier, early in her career. Converted from a collier, the Langley had a single, folding funnel to port and an opening to starboard, below the flight deck, for smoke disposal. Theoretically, both of them could be used, depending on the wind. Although the starboard opening had water sprays for cooling, it was not very useful nor successful. Later the carrier was refitted with two hinged stacks to port, which could be lowered for flight deck operations. Courtesy Jim Bulebush

ABOVE: Navy Lt. Cmdr. Harold T. Bartlett was flight commander of PN-10, a seaplane that made a non-stop flight from Hampton Roads to the Panama Canal Zone in 1926,traveling a distance of 2,061 miles. He also was honored with the Navy Cross for his service in World War I. Courtesy Hampton Roads Naval Museum

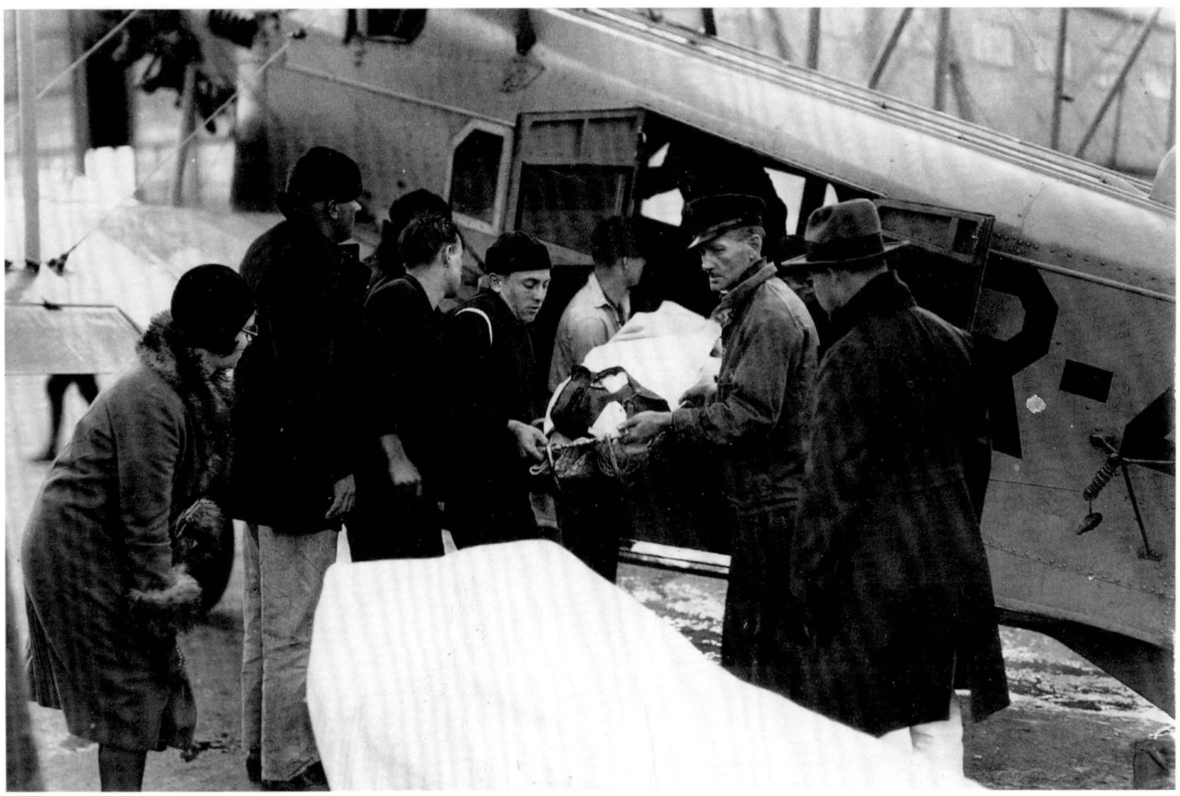

LEFT: An injured man is flown by a Norfolk Naval Air Station ambulance plane, a Loening OL-2, circa 1925. The Loening Aeronautical Engineering Company built this series of Army and Navy amphibians, which first appeared in 1923. The planes were used to pick up emergency patients from the Outer Banks and the Eastern Shore before bridges and tunnels were constructed. Courtesy Hampton Roads Naval Museum

ABOVE: Senior officers walk the flight line at Norfolk Naval Air Station, inspecting FB-5s belonging to Fighting Squadron 1B, 1927. Courtesy Official Navy Photo, Amy Waters Yarsinske Collection

RIGHT: On April 23, 1927, in Hampton Roads, Lt. Stephen W. Callaway set a world record for Class C seaplanes carrying a load of 1,102 pounds flying the Vought O2U pictured here. He attained an average speed of 147.263 miles per hour for 100 kilometers (a little over 62 miles). Courtesy The Amy Waters Yarsinske Collection

ABOVE: A 1928 fleet exercise off the Virginia coast involving the airship USS Los Angeles (ZR 3) accompanied by the USS Patoka (AO 9, foreground, below the airship), USS Lexington (CV 2, the aircraft carrier in the center of the picture) and USS Saratoga (CV 3, upper left, above the Lexington). Courtesy National Archives and Records Administration Photo, Amy Waters Yarsinske Collection

LEFT: Lt. Apollo Soucek walks away from his Wright Apache after establishing a new world record altitude mark, 38,560 feet, for Class C seaplanes on June 4, 1929. Courtesy Official Navy Photo, Amy Waters Yarsinske Collection

BOTTOM LEFT: Torpedo Squadron 1's mascot is ready to go flying in this circa 1929 photograph. This was the Navy's first torpedo squadron, formed in Yorktown in 1919. Courtesy Official Navy Photo, Amy Waters Yarsinske Collection

BELOW: An O2U-2 Corsair belonging to Scouting Squadron 3B takes off from the USS Lexington, 1929. Courtesy Official Navy Photo, Amy Waters Yarsinske Collection

ABOVE: The aircraft carrier USS Saratoga (CV 3) departs the Norfolk Navy Yard (now Norfolk Naval Shipyard) and passes downtown Norfolk, circa 1930. Courtesy Official US Navy Photo, NAS Oceana archives

RIGHT: The USS Langley, 1930. Her Navy career continued for another 12 years, until she was sunk in combat in the South Pacific during World War II. The ship received the Navy Unit Commendation. Courtesy Official US Navy Photo, Virginian-Pilot archives

LEFT: Sailors participate in a Memorial Day parade on Granby Street in Norfolk, 1932. Courtesy Sargeant Memorial Collection, Norfolk Public Library

FAR LEFT: A plane takes off the deck of the aircraft carrier USS Saratoga (CV 3) during a review by President Herbert Hoover off the Virginia coast, 1930. Courtesy Sargeant Memorial Collection, Norfolk Public Library

BOTTOM LEFT: Airplanes and Navy personnel on the Saratoga, 1930. Courtesy Sargeant Memorial Collection, Norfolk Public Library

ABOVE: Capt. Kenneth Whiting, commandant of Norfolk Naval Air Station, middle; and pilot Lt. Alfred M. Pride, right, inspect the Navy's new Autogiro aircraft, 1931. Purchased from the Pitcairn Aircraft Company, it was the Navy's first rotary-winged aircraft. Courtesy Sargeant Memorial Collection, Norfolk Public Library

RIGHT: A Martin BM-2 from Torpedo Squadron 1 skirts the Chesapeake Bay in 1932. The aircraft could handle a torpedo as well as a 1,000-pound bomb - the first Navy dive bomber that could handle such a large payload. Courtesy Official US Navy Photo, Amy Waters Yarsinske Collection

BELOW: The members of Patrol Squadron 10, which set a record for nonstop formation transpacific flight, from San Francisco to Pearl Harbor in 24 hours in 1934. Fourth from left is Jesse A. Fairley, who went on to become the first officer-in-charge of Oceana Naval Auxiliary Air Station in 1943. Courtesy Melba Carter

ABOVE: Early aircraft carriers, clockwise from bottom, USS Lexington (CV 2), USS Ranger (CV 4), USS Langley (CV 1) and USS Saratoga (CV 3), circa 1935.
Courtesy Clara W. Strehlow

LEFT: Capt. Charles W. Fox and his family at Norfolk Naval Air Station, May 9, 1934.
Courtesy Sargeant Memorial Collection, Norfolk Public Library

BOTTOM LEFT: A shellback ceremony, commemorating sailors' first crossing of the Equator, aboard the Saratoga in the 1930s. Courtesy Clara W. Strehlow

BELOW: Rudolph Wendler atop a torpedo bomber, Norfolk Naval Air Station, circa 1935.
Courtesy Clara W. Strehlow

ABOVE: A naval fighter plane aboard the USS Saratoga (CV 3) being brought up to the flight deck in the 1930s.
Courtesy Sargeant Memorial Collection, Norfolk Public Library

TOP RIGHT: Aerial view of Chambers Field, Norfolk Naval Air Station, 1939.
Courtesy Hampton Roads Naval Museum

RIGHT: A Loening amphibian ambulance lands in Norfolk after a flight from Hatteras in the 1930s. The Navy designated the aircraft XHL-1.
Courtesy Hampton Roads Naval Museum

LEFT: The flight deck of USS Ranger (CV 4) is loaded with planes, 1939. Courtesy Official US Navy Photo, NAS Oceana archives

FAR LEFT: The USS Yorktown (CV 5) at the pier, 1938. Courtesy Virginian-Pilot archives

BOTTOM LEFT: Sailors reunite with their loved ones, 1939. Courtesy Sargeant Memorial Collection, Norfolk Public Library

ABOVE: A crowd waves as the USS Enterprise (CV 5) departs, 1939. Courtesy Sargeant Memorial Collection, Norfolk Public Library

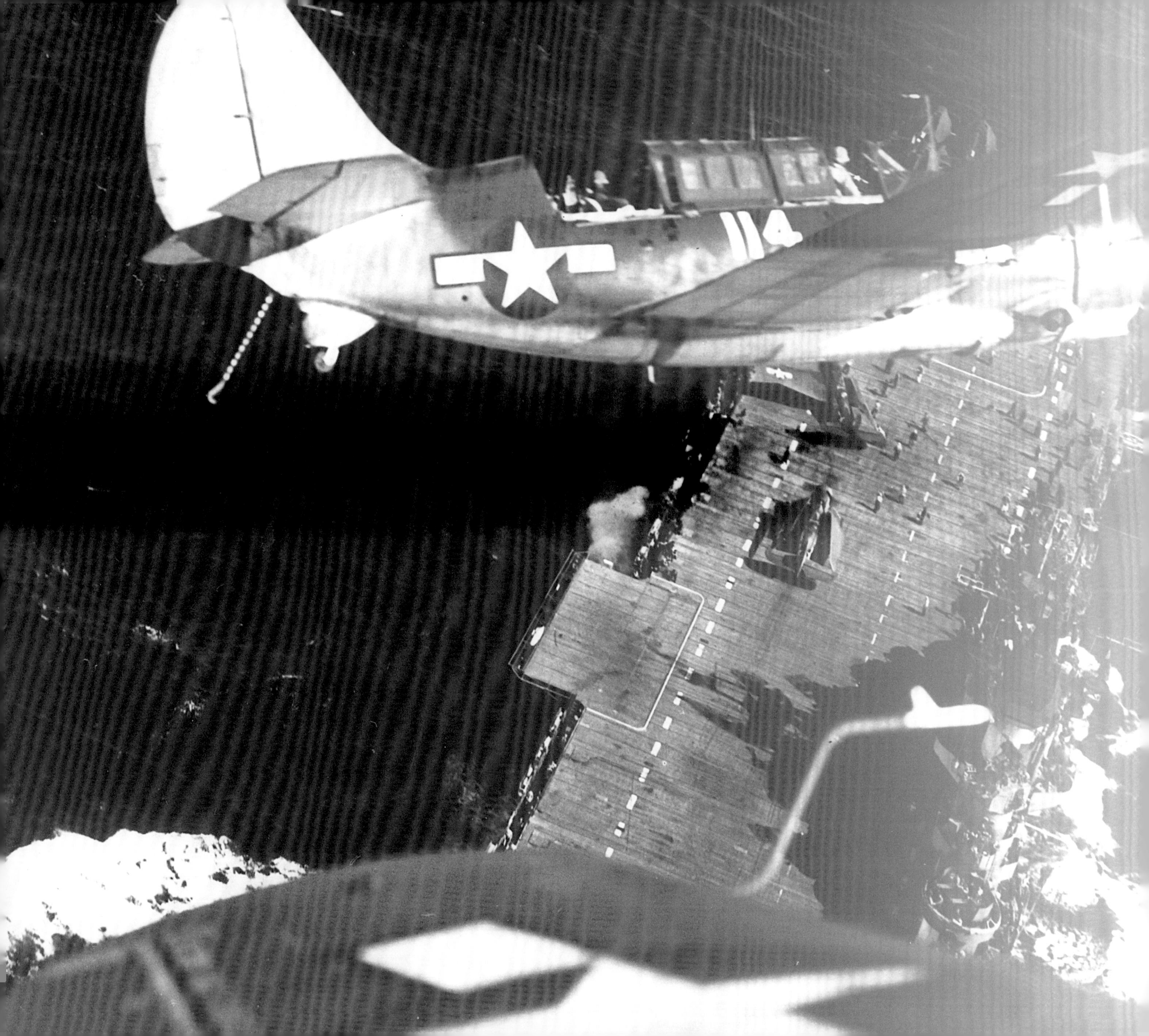

CHAPTER TWO

The 1940s

SAILORS' FEARFUL FAMILIES WAITING FOR word of casualties. Long-distance phone lines overwhelmed by frantic calls. Crowds in the streets and police on guard for sabotage. The newspaper headlines of December 8, 1941, show the stunned, sudden transformation of Hampton Roads into a region at war.

The tens of thousands of military personnel and wartime workers who poured into the area would change its landscape. About a year before the Pearl Harbor attack, the Navy had purchased 329 acres in the remote swamps of Princess Anne County for a small airfield that opened in November 1941. By 1943, it had grown to house nearly 1,000 men and been designated Oceana Naval Auxiliary Air Station.

Norfolk saw its own building boom, with new piers at the Naval Operating Base and hangars, runways and ramps to serve sea- and land-based aircraft at the Naval Air Station. By December 1942, the base had stopped training recruits in order to focus on advanced training for those going directly to the fleet.

Newport News Shipbuilding began constructing aircraft carriers at a breakneck pace – less than 18 months for the USS Yorktown (CV 10), which replaced a ship of the same name sunk at Midway.

As sailors and pilots deployed from Norfolk were fighting and dying overseas, tragedy struck at home. In September 1943, trailers full of ammunition exploded at Norfolk Naval Air Station, killing 40 in one of the worst Navy non-combat disasters of the war.

Though peace had come by the second half of the decade, the drive continued for faster, more powerful aircraft and bigger carriers to support them. By the decade's end, sleek new jets were launching from the decks of ships named in homage to the war's great battles.

OPPOSITE: An SB2C Helldiver returns from a strike against Japanese shipping, 1945. Courtesy National Archives and Records Administration, Amy Waters Yarsinske Collection

ABOVE: Seaplane control tower, Norfolk Naval Air Station, 1941. Courtesy Hampton Roads Naval Museum

BELOW: Interior of a PBM-3 Mariner patrol bomber in flight over Norfolk, 1942. Courtesy Hampton Roads Naval Museum

ABOVE: The USS Yorktown (CV 5) at the pier in Norfolk, circa 1938. Built at the Newport News Shipbuilding & Drydock Co., the carrier was commissioned September 30, 1937. Courtesy Virginian-Pilot archives

TOP: Aerial view of East Field, Norfolk Naval Air Station, 1941. Courtesy Hampton Roads Naval Museum

U. S. AND JAPAN AT WAR

Norfolk Virginian-Pilot

Norfolk, Virginia, Monday, December 8, 1941

Vol. CLXIII. No. 69

Three Cents the Copy

Japan Attacks Hawaii And Philippines

Norfolk Stunned By News; Japanese Seized

Anxiety Hits Kin of Navy; VPF 'Alert'

Traffic Jams Before Newspaper Office; 40,000 Extras Sold

Telephones Jangle

Long-Distance Overburdened; Police Told to Watch for Sabotage

Where the Twain Shall Meet

Vast Damage Rumored on Two Islands

Manila Dispatch Denies Axis Claim of Havoc in Islands

American and British Holdings in Far East Under Deadly Assault

Honolulu Blasted

Governor of Hawaii Tells of Heavy Loss of Life and Property

Roosevelt to Go Before Joint Session at 12:30 Today

Declaration Likely

Round-Up Begins

Nation's Chiefs Swing to Action As War Strikes

Japanese Herded, Leaves Cancelled by Army and Navy; Guards Placed

At Least 350 U. S. Soldiers Reported Killed in Hawaii

Singapore Blasted

Battleship Oklahoma Set Afire; Two Other Warcraft Attacked

Britain Makes Ready to Join In Pacific War

Parliament Summoned and Declaration of Hostilities Awaited

Japanese Seize Shanghai, Sink British Vessel

American Gunboat Wake Reported Captured, But Undamaged

News Awaited On Casualties

War and Navy Departments Urge Individual Queries Be Withheld

Japanese Bomb Big Naval Base Of Singapore

Air Attack Follows Landing Effected in Northern Malaya

All Harmony

Wheeler, Fish Call for Unity Against Japs

Japan Declares War on U. S., Great Britain

Anzac Clipper Safe in Hawaii

Canada Enters War on Japan

U. S. Battleship Sunk, Japanese Radio Asserts

On Inside Pages

15 DAYS TO SHOP

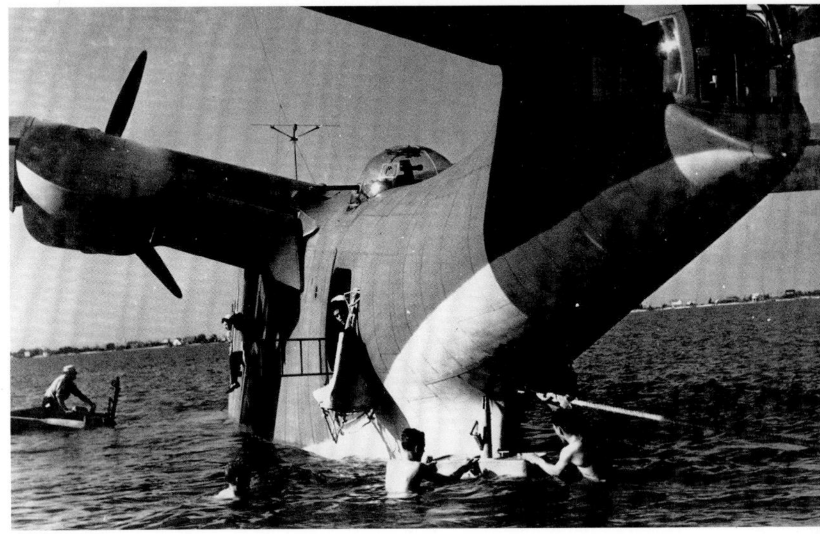

ABOVE: A PBM-3 Mariner anchored off Norfolk Naval Air Station, 1942. Courtesy Hampton Roads Naval Museum

LEFT: A day after the Pearl Harbor attack, the front page was already looking to the battles ahead in the Pacific. Courtesy Virginian-Pilot archives

BELOW: A B-25B bomber lifts off the deck of the USS Hornet (CV 8) in April 1942 as part of Doolittle's Raid - the first air raid on the Japanese homeland after Pearl Harbor. Courtesy Official US Navy Photo, Amy Waters Yarsinske Collection

ABOVE & TOP RIGHT: Advance Base 'A' Training Unit, hangar no. 2, Breezy Point seaplane operating area at Norfolk Naval Air Station, 1942. The working and sleeping quarters combined for 1,100 men.
Courtesy Hampton Roads Naval Museum

RIGHT: The first word of Doolittle's raid came from Tokyo radio stations, the Associated Press reported. Courtesy Virginian-Pilot archives

OPPOSITE: Army Lt. Col. James H. "Jimmy" Doolittle wires a Japanese medal to the fin of a 500-pound bomb on the deck of the USS Hornet before the raid on Tokyo. The bombers were craned aboard the carrier at Norfolk Naval Air Station.
Courtesy Official US Navy Photo, Amy Waters Yarsinske Collection

Weather Forecast
Virginia: Cooler Saturday. North Carolina: Slightly cooler in the interior Saturday. (Full U. S. Weather Bureau report on Page 2.)

Norfolk Virginian-Pilot

Vol. CLXV. No. 18 — Norfolk, Virginia, Saturday, April 18, 1942 — Three Cents the Copy

Tokyo Bombed, Japanese Radio Reports

Stimson Predicts Offensive Soon On Far Flung Fronts

Key Points in Distant Lands Must First Be Made Secure

Jap Fliers Downed

Australians Get at Least Four Raiders Over Port Moresby

Washington, April 17.—(AP) Secretary Stimson said today the Army would be ready for the offensive soon, no matter what difficulties might be encountered. At the same time he emphasized that Gen. Douglas MacArthur had "over-all strategic command" in the Southwest Pacific.

(Later in the day, news dispatches from Australia said MacArthur had received a directive from Washington clarifying his powers. Newsmen at U. S. Army headquarters in Australia were informed the directive was completely satisfactory.)

Without minimizing the obstacles and problems involved in preparing for offensive warfare of world-wide scope, the Secretary of War spoke both confidently and optimistically to newspapermen at a press conference.

Nearly Ready Now

"So far as the Army is concerned," he said, "we are getting pretty near the stage of being ready for an offensive, however difficult it may be.

Things are beginning to move

40 French Ships Under Swastika, Says Red Radio

London, April 17.—(AP)—The Moscow radio said tonight that "already 40 French warships have changed the tri-color for the Nazi swastika."

"Is this the writing on the wall?" the broadcaster asked in an English news commentary.

There was no elaboration of the statement. There have been reports that the Germans took over unfinished submarines and possibly some destroyers when they occupied France, and that they have since finished them themselves.

British Wreck Sources of Oil In West Burma

Six Thousand Wells in Flames; Yorkshiremen Fight Valiantly

New Delhi, India, April 17.—(AP) British sappers burned and blasted 6,000 west Burma oil wells today while a gallant battalion of the King's Own Yorkshire Light Infantry fought yet another superb delaying action against the Japanese until the wells were ablaze.

Then the vastly outnumbered

Step to Block Labor-Profits Bill Defeated

Committee Defers for 10 Days Vote On Vinson Measure

Lewis Under Fire

Mine Workers' Chief Accused of Attempting to Throttle Farmers

Washington, April 17.—(AP)—An attempt to block action by the House Naval Committee on War Profits and Labor Legislation for at least a month was defeated by a 9-to-9 vote today, demonstrating with new force the sharp cleavage of congressional opinion on the issues.

The delay was proposed by Representative Flaherty, Democrat, Massachusetts, and opposed by Chairman Vinson, Democrat, Georgia, who told the members, "These issues are so dominant that they are going to be prevailing in every congressional district by election time."

Flaherty urged delay on the grounds that President Roosevelt might work out some program disposing of the issues.

"We're all aware," he said, "of the possibility of executive action on all phases of this legislation. Negotiations to that end are now going on, and rejection of this bill might have a bad effect on those deliberations."

"And what effect would it have if we reported it up?" countered Vinson, drawing no reply.

By agreement, however, the committee decided to defer a vote for 10 days on Vinson's measure calling for a percentage limitation on war profits, suspension of the 40-hour week law, and preservation of the status-quo of unions in plants handling Army and Navy contracts.

Heaviest RAF Raids of War Harass Axis

Day-Long Forays Reach Heart of Industrial Germany

Opposition Slight

Only Two Planes Lost Out of 600 Hurled Against Foe, London Says

By Drew Middleton

London, April 17.—(AP)—Adding weight and distance to its massive daylight offensive, the RAF hurled 600 planes against German targets today in day-long raids extending from the Continental coast to Augsburg, in southern Germany.

The assaults were by far the heaviest of the war by British airmen, topping yesterday's far-flung, 400-plane attack, which had been the biggest to date.

Emphasizing the enormity to the aerial offensive was a daylight attack upon Augsburg, which involved a round trip of at least 1,000 miles right over the heart of industrial Germany.

Augsburg, a few miles northwest of Munich, is the site of a Messerschmitt plane factory, but an informed source said this establishment was not the target of today's raiders. This center was the target of night raids twice in August, 1940, but had not been mentioned in British air communiques since then.

Only Two Planes Lost

British air superiority over Western Europe was stressed by the Air Ministry announcement that only two out of the hundreds of planes used—one fighter and one bomber—were lost. It described the enemy opposition as "relatively slight."

The Air Ministry said at first that two fighters and a bomber were lost, but a later announce-

Schools and Hospitals Damaged, And Casualties Yet Unknown, Says Nippon; Washington Silent

Leahy Is Called Home for 'Consultation' As Welles Declines to Accept Vichy Note

Acting Secretary of State Denounces Puppet Government

By Wade Werner

Washington, April 17.—(AP) President Roosevelt today recalled Ambassador William D. Leahy from Vichy "for consultation" under circumstances indicating that this country has no intention of attempting to maintain normal relations with a collaborationist regime dominated by Pierre Laval.

By way of emphasizing this American rejection of Laval, Acting Secretary of State Sumner Welles let it be known that this Government views the new Vichy regime as a puppet government capable of sending puppet communications previously submitted to and approved by German authorities.

Due to Laval's Rise

Ambassador Leahy will take his departure as soon as Mrs. Leahy, who recently underwent

Airliner Shades Hide War Work From Travelers

Washington, April 17.—(AP)—Airline passengers, it was disclosed today, now are directed to pull down their window shades when their plane takes off or lands.

The idea is to prevent scrutiny of war industries, harbor defenses and the like in the vicinity of airports, it was explained.

Generally the shades must be drawn five minutes before landing, and five minutes after taking off, but the time varies according to the surroundings of the various airports. The regulations were adopted by the Air Line Transport Association, it was said, and the understanding was that the Army had suggested them.

Laval Forming New Cabinet After Mass Resignations

By Taylor Henry

Vichy, April 17.—(AP)—Adm. William D. Leahy, U. S. Ambassador to Unoccupied France since November 23, 1940, was called home for consultation tonight as pro-German Pierre Laval rounded out a new Cabinet to replace the Government which resigned en masse to Marshal Petain.

Late tonight, however, Laval postponed until tomorrow announcement of a final Cabinet list.

Official France had, in the interim, two leaders—the aged marshal as chief of the state who was forced to bow before German demands for increased collaboration and Adm. Jean Darlan, designated as Petain's successor and commander-in-chief of all the armed forces.

The remainder of the Govern-

(Continued on Page 3)

First Repayment for Pearl Harbor Apparently Undar Way

No Confirmation

Time Required for News of Long-Distance Attack, Officials Say

Tokyo, April 18 (from Japanese Broadcasts)—(AP)—The Tokyo region was raided by enemy planes this afternoon, a Japanese official announcement said today. The raiders came from several directions.

Three planes were declared to have been shot down in the raids, the first of the war on Japan's imperial homelands.

By William Smith White
Associated Press War Editor

The long-awaited bombing of Tokyo, the first installment of repayment for Pearl Harbor and the start of offensive warfare directly against the heart of Imperial Japan, apparently got under way at noon today, for the Tokyo radio announced that enemy planes had raided the city for the first time.

There was no immediate confirmation from the United

More Than 65,000 Taken By Japanese as Bataan Fell,

RIGHT: An early Navy twin-engine assault drone, the TDN-1, built in 1943. It was a precursor to the modern-day Predator drone. This drone was tested in Hampton Roads.
Courtesy Hampton Roads Naval Museum

BELOW: Rear Adm. Patrick N. L. Bellinger (far right), commander of Naval Air Force Atlantic, attends the ceremony establishing Oceana Naval Auxiliary Air Station on August 17, 1943. Lt. Jesse Fairley, the air station's first officer-in-charge, is standing center
Courtesy Official US Navy Photo, NAS Oceana archives

ABOVE: James Jethro "J.J." Gardner, right, of Norfolk, was a plane mechanic aboard the USS Hornet (CV 8) - one of four ships he served aboard in World War II that were sunk. He was sent to San Diego to help sell war bonds and posed for a photo with actor Henry Fonda at the Sherman Bar, a popular watering hole, April 1943. Courtesy James J. Gardner

Norfolk Virginian-Pilot

Vol. CLXV, No. 67 — Norfolk, Virginia, Monday, June 8, 1942

Great Battle Raging Off Midway Island

Food Control Board Headed By Wickard

Production and Use of Supplies in Hands of Committee

Shortages Possible

Secretary Asks Girls to Stay on Farms and Help With Work

A Statement to Our Readers and Advertisers

Following the temporary cessation of work by members of the Norfolk-Portsmouth Typographical Union No. 32 on the night of May 7, which forced an abridgement of the paper affected by this stoppage, there was a resumption of normal operations as well as a resumption of the wage-contract negotiations that were being amicably conducted during that period.

During the last four weeks conferences have been held between the publishers of the Norfolk papers on the one side, and a committee acting for the typographical union on the other, in an effort to reach an agreement on a new working contract to replace the contract which expired on April 30, 1942. Present at the most recent of these meetings was a representative of the International Typographical Union with which the local printers' union is affiliated.

On Wednesday, the president of the local printers' union notified the president of Norfolk Newspapers, Inc., that the union had voted to reject the publishers' offer for a composition of differences. Concurrently with this action, the members of the union voted to ask the International Typographical Union for permission to take a strike vote. The publishers are informed that there is a possibilty that the governing body of the International Union may sanction the strike vote. In that event it would lie within the power of the local union after voting to strike, to cease work without further notice.

In an effort to avert a suspension of newspaper service in this critical national period and to bring all differences to a prompt reconciliation, the publisher has requested the president of the printers' union to join him in submitting the questions at issue to arbitration.

We hope that the arbitration request will be acceded to. That is the course that commends itself in these times above all others where agreement cannot be reached by negotiation, and is the course recommended by the National War Labor Board and the President of the United States.

Should the arbitration proposal be rejected, there remains the possibility that a strike may force the daily newspapers of Norfolk either to suspend publication during its pendency or to publish in severely restricted form.

The purpose of this statement is to apprise our readers and advertisers of the state of affairs just described. Without at this time discussing the substance of the differences between the publishers and the printers, we think it is fair to state to our readers and advertisers that, in our opinion, we have left

Capital Split On Issue of Gas, Rubber

Hundred Congressmen Fight Rationing in Noisy Session

Roosevelt Confers

No Hope Civilians Can Get Tires for Years, Says Eastman

Warning Given

Service Pay Base of $46 Is Approved

Senate-House Conferees Reach Agreement; Fight Continues

Dates From June 1

Benefits for Dependents of Civilians Given Committee Approval

Fate of Japanese Plans To Raid U. S. Mainland Is Believed in Balance

Shell Plant Blast Kills 21; 33 Missing

Production Halts in Only One Unit of Huge Elwood Works

Promoted

British Airmen Raid Nazi-Held Ports, Plants

American-Made Boston Bombers Hit Ostend and Other Cities

Marines, Naval Units and Air Forces Defend Outpost

Enemy Warcraft Hit

Damage Inflicted by Foe Small; Report of Pearl Harbor Attack Denied

ABOVE: A view of Norfolk Naval Air Station on September 7, 1943.
Courtesy Hampton Roads Naval Museum

TOP LEFT: The USS Yorktown (CV 10) was commissioned April 15, 1943, less than one year after her namesake, designated CV 5, was sunk at the Battle of Midway. The aircraft carrier was built in just 16 and a half months at Newport News Shipbuilding and was in combat by August 1943.
Courtesy Official US Navy Photo, Virginian-Pilot archives

LEFT: The June 1942 Battle of Midway quickly proved to be the turning point of the war in the Pacific. American carriers and shore-based patrol bombers sunk four Japanese carriers and destroyed or damaged numerous other enemy ships.
Courtesy Virginian-Pilot archives

NORFOLK LEDGER-DISPATCH

HOME EDITION

NORFOLK, VA., FRIDAY AFTERNOON, SEPTEMBER 17, 1943

17 DIE IN NAVY BLAST HERE; 257 ARE REPORTED INJURED

British 8th, U.S. 5th Make Juncture

Plans Drafted For New Blows At Reich And Japan, Says FDR

$128,123,000,000 Cost Of War To U.S. To Date

Three Armies Close In On Krasnograd

Explosion At Air Station Heard More Than 20 Miles

Italian Railway Links Blasted

Lae Tottering; Fall Imminent

War-in-Brief

Eight Indicted As Nazi Spies

New Protest On Draft Made

Six Miners Escape Blast

Germans Force Italian Workmen Into Slave Labor

Critical Milk Shortage Revives Subsidy Issue

ABOVE: In one of the worst non-combat disasters of World War II, several trailers full of depth charges exploded at Norfolk Naval Air Station on September 17, 1943, injuring about 400 and killing 40, including Seaman 2nd Class Elizabeth Korensky, the first WAVE to die in the line of duty. Courtesy Virginian-Pilot archives

TOP RIGHT & RIGHT: Sifting through the wreckage and surveying the damage after the explosion. Courtesy Hampton Roads Naval Museum

ABOVE, TOP LEFT, BELOW: Cleanup efforts in the aftermath of the ammunition explosion on September 17, 1943. Courtesy Hampton Roads Naval Museum

LEFT: Navy nurses were quick to respond on the scene of the explosion. Courtesy Sargeant Memorial Collection, Norfolk Public Library

ABOVE: Sailors in the naval air station's supply scrap yard, 1944. Courtesy Hampton Roads Naval Museum

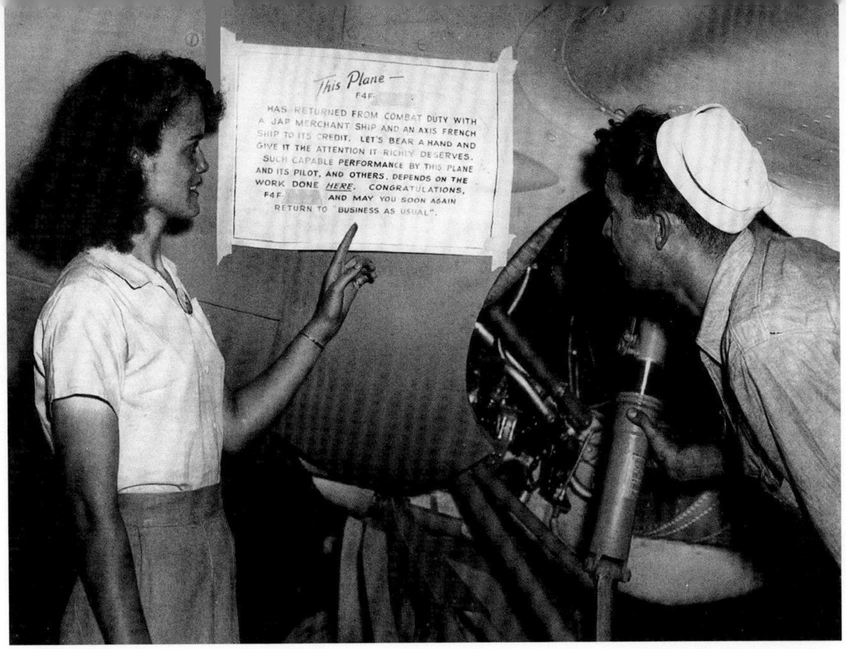

ABOVE: Francis A. Greene plays his clarinet during a free moment at the naval air station, circa 1944. Courtesy Donald F. Greene

TOP LEFT: An F4F Wildcat at Norfolk Naval Air Station's Assembly and Repair Department. The sign on its fuselage reads, "This plane has returned from combat duty with a Jap merchant ship and an Axis French Ship to its credit. Let's bear a hand and give it the attention it richly deserves. Such capable performance by this plane and its pilot, and others, depends on the work done here. Congratulations F4F, and may you soon again return to 'business as usual.'" Courtesy Hampton Roads Naval Museum

BOTTOM LEFT: Rear Adm. Gerald F. Bogan on the flag bridge of the USS Bunker Hill (CV-17). Two kamikaze pilots struck the aircraft carrier in May 1945, killing nearly 400. Courtesy National Archives and Records Administration Photo, Amy Waters Yarsinske Collection

BELOW: Recovery of an F6F Hellcat at Gate 3, Norfolk Naval Air Station, 1944. Courtesy Donald F. Greene

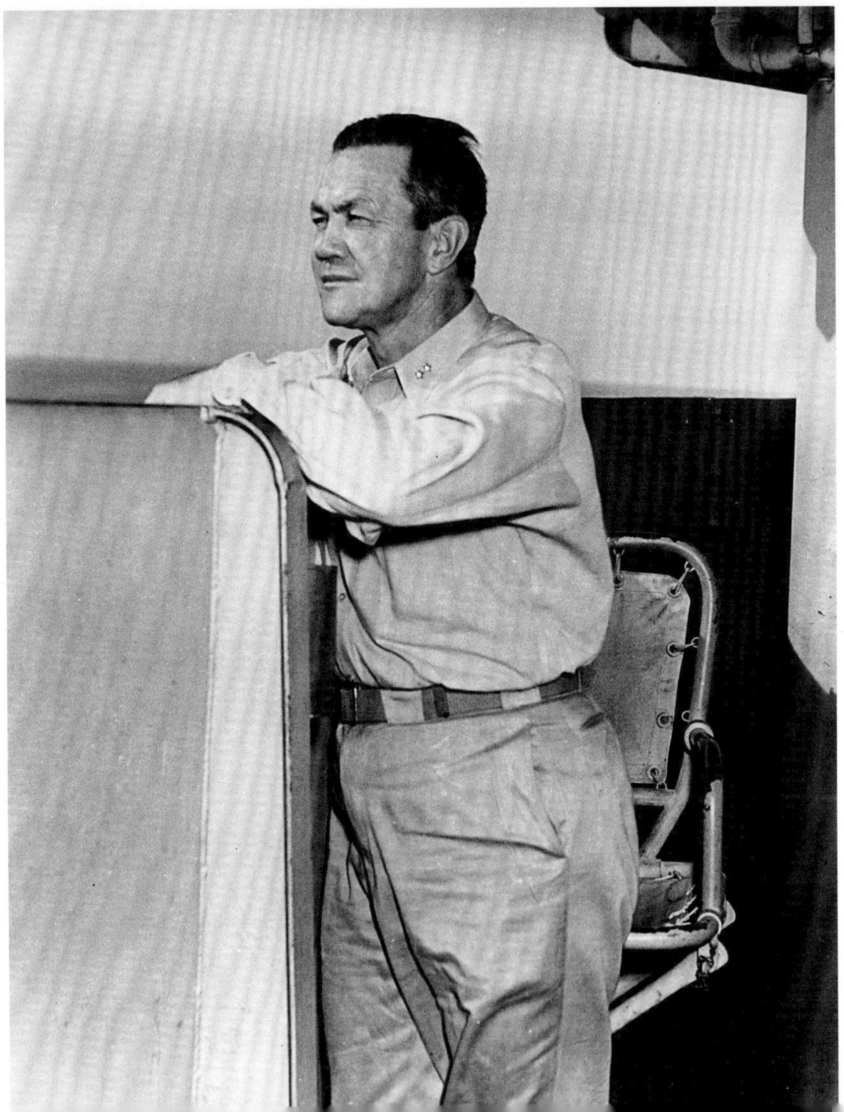

ABOVE: The USS Ranger (CV 4) was based in the Atlantic during World War II, where it played a decisive role in Operation Torch, the 1942 invasion of North Africa. The carrier also ferried hundreds of Army fighter planes overseas, including the P-38 Lightnings shown here, 1944. Courtesy Official US Navy Photo, Amy Waters Yarsinske Collection

ABOVE: Cmdr. David McCampbell with his plane captain Chet Owens in November 1944. McCampbell's F6F Hellcat fighter is adorned with all 34 of his victories in air combat. The leading Navy ace at the time, he was awarded the Medal of Honor for his actions during the battles of Philippine Sea and Leyte Gulf. Courtesy Official US Navy Photo, Amy Waters Yarsinske Collection

RIGHT: Pilots from Torpedo Squadron 51 return to the USS San Jacinto (CVL 30) after a strike on the Japanese carrier fleet during the Battle of Leyte Gulf, October 1944. A TBF-1 Avenger is in the background. The squadron trained at Norfolk and Oceana before it departed for the Pacific. Its pilots included George H.W. Bush, who went on to become President. Courtesy Official US Navy Photo, Amy Waters Yarsinske Collection

LEFT: Midshipmen in their dress whites on the deck of the carrier USS Lake Champlain (CV 39), circa 1945.

Courtesy Sargeant Memorial Collection, Norfolk Public Library

BOTTOM LEFT: The Battle of Leyte Gulf, October 23-25, 1944, was perhaps the largest naval encounter ever. It severely reduced the Japanese Navy's ranks.

Courtesy Virginian-Pilot archives

BELOW: Navy personnel aboard the USS Lexington (CV 16), circa 1945.

Courtesy Sargeant Memorial Collection, Norfolk Public Library

NORFOLK LEDGER-DISPATCH

THE WEATHER

VOL. CXXXII—NO. 101 NORFOLK, VA., THURSDAY AFTERNOON, OCTOBER 26, 1944 28 PAGES—PRICE FIVE CENTS

EDITION

26 MAJOR JAPANESE SHIPS KNOCKED OUT

War May Be Shortened By Months, Says Nimitz

Berlin Reports Allied Landings On Bevelaud

Canadians Close in On Last Big German Guns Blocking Antwerp

By WILLIAM FRYE

London, Oct. 26 (AP)—The German High Command announced today that Allied forces, striking to clear the last Nazi guns blocking Allied convoys from Antwerp, had landed on the south shore of Beveland island.

"Countermeasures are in progress," the communique said, indicating the amphibious forces already had made good a bridgehead on the north side of the Schelde estuary.

Earlier Berlin broadcasts that Walcheren, another island north of the estuary, had been invaded by Allied land and sea forces were not borne out by the communiques.

Land Under Protection of Fog

The landing apparently was made under the protection of a dense morning fog which swept in from the North sea and came as Allied forces, fighting through hip-deep water, forged slowly ahead to clear the sea approaches to Antwerp.

Field Marshal Sir Bernard L. Montgomery's pincers tightened relentlessly on Germans clinging to the water-logged deltas and islands of Southwest Holland, with the enemy giving ground all the way from Bergen Op Zoom to 'S Hertogenbosch.

Driving in from the east on 11,000 Germans trapped on Walcheren, the Canadians fought through thick fog and traversed half the length of the two-mile-wide, 10-mile-long neck to the connecting island of South Beveland. They captured Rilland on the way.

Advance On Single Roads

The progress was along single roads raised over wastes of water left by flooding of the islands.

South of estuary, the Canadians were fighting in the streets of Oostburg, four miles southeast of Cadzand, where the enemy's last guns block the estuary from the south. They also recaptured Fort Frederik Hendrik and reached the outskirts of Groede, four and one-half miles northeast of Cadzand.

Behind the advance into South Beveland they captured the village of Dorp Op Pindorp, four miles east of the enemy coastal anchor of Ber-

Red Troops Drive Deeper Into Norway From Kirkenes

Russians in Supreme Effort to Annihilate 130,000 Germans on Far Northern Front

Moscow, Oct. 26 (AP)—Russian troops drove deeper into Norway today beyond captured Kirkenes in Arctic storms which tumbled the temperature several degrees below zero, in a supreme effort to annihilate up to 130,000 Germans being herded before Red and Finnish forces.

In local engagements just north of ravaged Warsaw, the Kremlin at midnight announced the capture of a dozen satellite towns on the east side of the wide and swift Vistula river which bends sharply west northwest just above the ravaged city.

The pace of the Red Army offensive in East Prussia slackened before ferocious German resistance, explained in part by the dependency of Berlin and other Nazi cities on foods produced on East Prussian farms.

All northern Transylvania was behind Russian lines in the Balkans and Romania was cleared of hostile Germans. In Czechoslovakia, the Russians virtually isolated the big highway town of Mukacevo, pushing within nine miles of the city.

Threat to Warsaw Ups

Clearing of the towns on the northern flank of Warsaw increased the threat of capture of the Polish capital.

The Soviet communique described the Warsaw successes as "battles of local importance," but they seemed the first real sign of German weakness in central Poland since the summer Red Army offensive faded out at the end of overstretched supply lines. (The Germans yesterday reported the Russians had launched a powerful offensive north of Warsaw.)

If Rokossovsky has built up a new punch in recent weeks—and there seems good reason to believe he has done just that—momentous operations may be developing that will overshadow anything else on the eastern front.

While the capture of Warsaw itself would have only psychological value now, a major breakthrough in the area would represent the double threat of an advance straight at Berlin as well as northward to East Prussia.

Meanwhile, the invasion of East Prussia ran into a storm of fanatical Nazi counterattacks around Gumbinnen and Goldap, and while the advance continued, it brought gains of less than two miles on a

ARCHBISHOP OF CANTERBURY

Canterbury Heart Victim

London, Oct. 26 (AP)—The Archbishop of Canterbury, the Most Rev. William Temple, died today after a heart attack. The Primate of All England was 63 and had been enthroned in bomb-pocked old Canterbury cathedral since April 23, 1943.

His death and that of Princess Beatrice, which occurred a few hours earlier, plunged England into mourning.

Dr. Temple was the son of an archbishop of Canterbury and they were the only father and son ever to hold the highest office in the Church of England. The father was the late Dr. Frederick Temple, who was archbishop during the reign of Edward VII.

Dr. William Temple succeeded Dr. Cosmo Gordon Lang, who resigned at 78 with the request that a younger man be placed in the office. In 1929, Dr. Temple had become Archbishop of York, second highest post in the Church of Eng-

Yanks Slam Shut South End Of Leyte Trap

Entire American Front Moves Ahead Through Mud and Rain

By MURLIN SPENCER

General MacArthur's Headquarters, Leyte, Philippines, Oct. 26 (AP)—Mud-caked Yanks today slammed shut the southern door of strategic Leyte valley, a potential Japanese death trap, and tightened their stranglehold on Nipponese communications with adjacent islands.

As naval and air units beat off the Japanese fleet threat, the entire American land front moved forward through mud and rain, beneath the cover of supporting aircraft and growing artillery fire.

The southernmost of three columns driving into Leyte valley, where Japanese defenders are being crowded, captured Burauen, a key road juncture nine miles inland from Dulag. Armor-led infantrymen of the 24th corps broke into the town, after taking nearby San Pablo and its cluster of airfields.

Landing On Samar Confirmed

On the extreme northern front, the 10th Corps completed shore-to-shore movements which seized the southern coast of nearby Samar island and shot American units to the northern tip of Leyte. The landing on Samar, fifth Philippine island to be invaded, was reported Monday and confirmed in today's communique.

The maneuver gave the Americans control of San Juanico strait, the narrow, crooked waterway dividing Leyte and Samar. The main Japanese supply line for supplies and reinforcements from the north thus has been severed.

Liberation troops have swept 20 miles northward of Tacloban, the communique reported. Apparently they will curve around the northern tip of Leyte to drive a wedge in Japanese communications with Masbate, main island to the northwest.

Jap Forces Disintegrating

Japanese forces in this sector "are disintegrating," the announcement added.

Carrier planes turned back a large-scale Japanese daylight assault Tuesday on American shipping in Leyte gulf. Fifty-three Nipponese planes were shot down in dogfights, and antiaircraft fire destroyed three more.

Ground forces "easily repulsed" a

THREE-WAY PHILIPPINES BATTLE—Numerals indicate Japanese naval forces sent toward the Philippines in moves leading to three-pronged naval battle which resulted in 26 major Nip warships being sunk, probably sunk or damaged. Arrows show U. S. and Jap moves.

Kincaid Meets Jap Audacity Saves MacArthur's Invaders

By SPENCER DAVIS

Gen. MacArthur's Headquarters, Philippines, Oct. 26 (Delayed)—(AP) Had it not been for Vice Adm. Thomas Kinkaid's decision to meet audacity with audacity, Gen. Douglas MacArthur's Leyte beachhead and supporting vessels in San Pedro harbor might have been under the fire of Japanese battleship guns this morning.

American PT boats sighted the enemy shortly after midnight yesterday and roared in for a brave, but unsuccessful, attack. Destroyers followed in with a closely pressed torpedo attack which sent Japanese battleships reeling under the impact of direct hits.

Battleships under command of Rear Adm. Jesse Oldendorf, some of which had been resurrected from the mud of Pearl Harbor,

Enemy Fleet Fades As Major Obstacle To Invasion Of Asia

Report of Losses Suffered By Foe Still Incomplete

By CHARLES H. McMURTRY

U. S. Pacific Fleet Headquarters, Pearl Harbor, Oct. 26 (AP) — More than 26 major Japanese warships were sunk or heavily damaged in the three-pronged battle of the Philippines, incomplete reports on the most crushing defeat ever administered the modern Nipponese Navy disclosed today.

Wreckage strewn waters marked the possible end of the imperial fleet as a major obstacle in the path to the coasts of Japan and China. The defeat of the Jap fleet may have shortened the Pacific war by months, said Adm. Chester W. Nimitz.

Combined reports of Adm. Nimitz and Gen. Douglas MacArthur on the three major engagements involving the bulk of the Japanese fleet listed these known results:

Eight Japanese ships sunk—two carriers, one battleship, five cruisers.

Two Battleships Probably Sunk

Three probably sunk — two battleships, one carrier.

More than 15 damaged — Seven battleships, four cruisers, four destroyers and "several" more destroyers.

Both commanders made it clear that their final reports will show heavier Japanese losses, particularly in ships sunk. Today's combined total compared with Nimitz' first report of the three way action listing one carrier sunk, two damaged, and five or six battleships and a cruiser torpedoed. All of these, with the possible exception of two battleships, are included in the combined total.

The crushing three-fold defeat stripped Emperor Hirohito's Navy of the greater part of its known carrier, battleship and cruiser power, necessary to engage the U. S. fleet in another major action or seriously check its further incursions into the western Pacific.

Invasion Hung In Balance

During the three-day battles, the fate of MacArthur's invasion hung in the balance. This was the touchstone that brought the Japanese fleet out of hiding and precipitated

U. S. Seventh Saves Leyte Invasion in Hour Of Greatest Peril

By C. YATES McDANIEL
Associated Press War Correspondent

General MacArthur's Headquarters, Philippines, Oct. 26 (AP)—Saving America's Leyte invasion in its hour of direct peril, the outnumbered but never outgamed Seventh Fleet sent carrier planes today in pursuit of two defeated Japanese naval armadas while U. S. Third Fleet planes bombed and torpedoed a third enemy warship force south of Formosa.

Already, with many reports yet to be heard, it is known that a Japanese large carrier, a big battleship, several cruisers and destroyers have been sunk, two carriers, four battleships and other warships damaged.

It is an historic victory. But there was one hour Wednesday morning as two Nipponese battleship-led groups emerged on the east side of the Philippines and approached Leyte gulf when the fate of Gen. Douglas MacArthur and his men hung in the balance.

Daring Decision By Kinkaid

Then it was that Vice Adm. Thomas C. Kinkaid made a daring decision. Despite numerical inferiority as a group, he boldly divided his Seventh Fleet and an Australian squadron, sent them in different directions and won both battles.

Closing in for the first gun battles in more than a year with the Japanese and backed by carrier planes, the Seventh:

Sank a 29,300-ton enemy battleship;

Sank as yet not totaled number of cruisers and destroyers;

Damaged at least three more cruisers and several destroyers;

Lost an escort carrier and several patrol torpedo boats;

(Continued On Page 26)

Tokyo Admits Battleship Lost

By ASSOCIATED PRESS

ABOVE: Christening of the USS Midway (CVB 41), the first of a three-ship class of 45,000-ton aircraft carriers, at Newport News Shipbuilding, March 1945. Courtesy Virginian-Pilot archives

LEFT: The light cruiser USS Santa Fe (CL 60) comes to the aid of the aircraft carrier USS Franklin (CV 13), after it was set afire by a Japanese dive bomber, killing 724 sailors, March 1945. Courtesy Official US Navy Photo, Virginian-Pilot archives

RIGHT: Germany's surrender was met with fireworks in Paris, celebrations in London, and, by the mother of Gen. Dwight D. Eisenhower, with the hope that "Dwight will be home soon." Courtesy Virginian-Pilot archives

BELOW: Reon G. Hillegass Jr. of Norfolk was an eighteen-year-old sailor on board the USS Franklin (CV 13) when it was hit on March 19, 1945. His ordeal shaped his life. "I grew up that day," he said later. "It changed my life forever. It made me a more determined person, more determined to take on a job and get it done, live my life like I want to live it, and enjoy myself doing it all." Courtesy The Amy Waters Yarsinske Collection

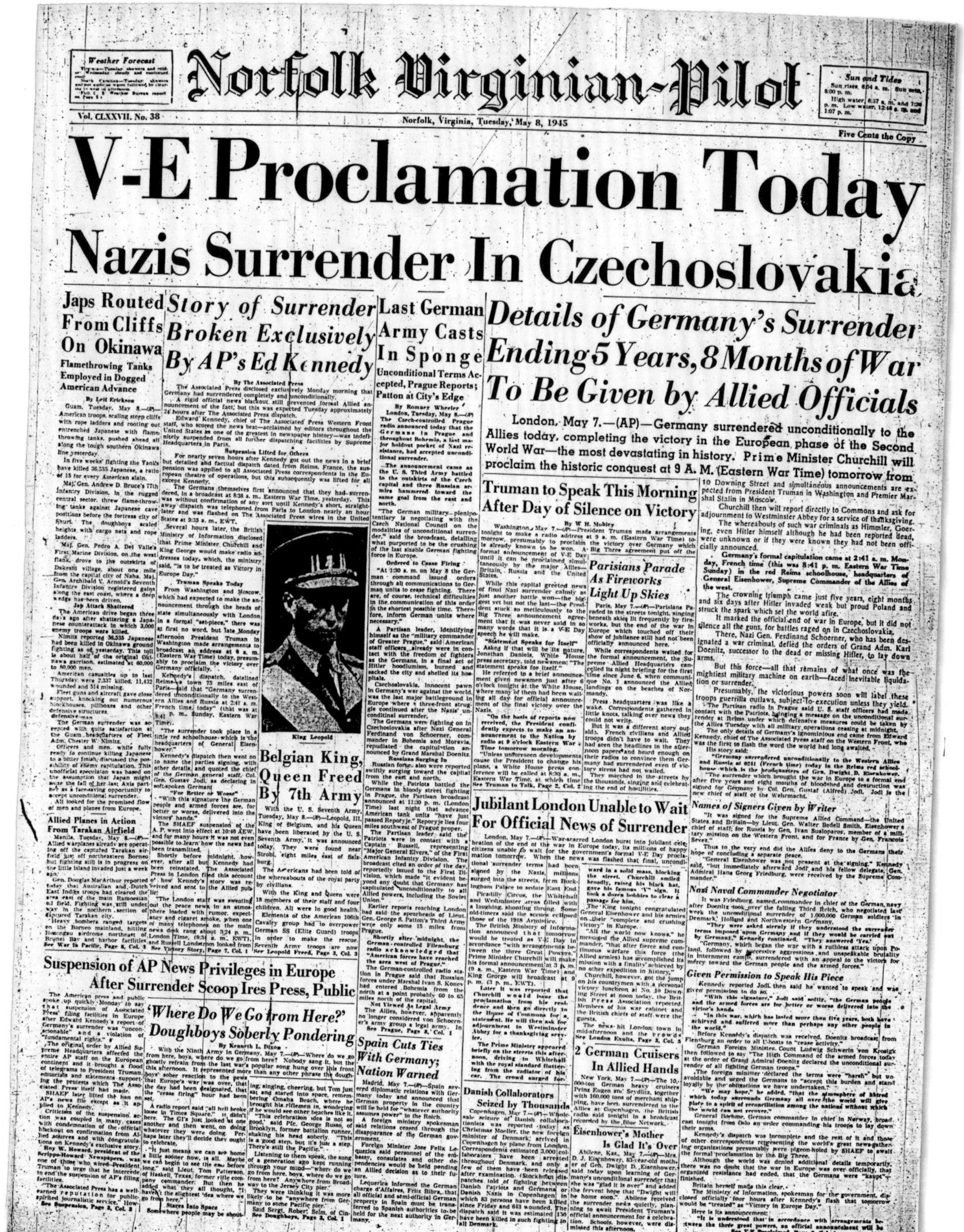

Norfolk Virginian-Pilot

Vol. CLXXVII. No. 38 — Norfolk, Virginia, Tuesday, May 8, 1945 — Five Cents the Copy

V-E Proclamation Today

Nazis Surrender In Czechoslovakia

Japs Routed From Cliffs On Okinawa

Flamethrowing Tanks Employed in Dogged American Advance

Story of Surrender Broken Exclusively By AP's Ed Kennedy

Last German Army Casts In Sponge

Unconditional Terms Accepted, Prague Reports; Patton at City's Edge

Details of Germany's Surrender Ending 5 Years, 8 Months of War To Be Given by Allied Officials

London, May 7.—(AP)—Germany surrendered unconditionally to the Allies today, completing the victory in the European phase of the Second World War—the most devastating in history. Prime Minister Churchill will proclaim the historic conquest at 9 A. M. (Eastern War Time) tomorrow from

Truman to Speak This Morning After Day of Silence on Victory

Parisians Parade As Fireworks Light Up Skies

King Leopold

Belgian King, Queen Freed By 7th Army

Jubilant London Unable to Wait For Official News of Surrender

Suspension of AP News Privileges in Europe After Surrender Scoop Ires Press, Public

'Where Do We Go from Here?' Doughboys Soberly Pondering

Spain Cuts Ties With Germany; Nation Warned

Danish Collaborators Seized by Thousands

Eisenhower's Mother Is Glad It's Over

2 German Cruisers In Allied Hands

TOP LEFT: Foreign naval attaches visit Norfolk Naval Operating Base, 1945. Pictured here with U.S. Navy officers who acted as guides, from left, are; First row, Rear Adm. Ignacio Garcia Jurado, Mexico; Rear Adm. J.L. Maijer Renneft, Netherlands; Gen. Shang- Chen, China; Rear Adm. David LeBreton, USN; Rear Adm. P. T. Mar, China; Rear Adm. Alberto D. Brunet, Argentina; Rear Adm. M.G. Slarrow, USN. Second row, Capt. A.F. Belikhow, USSR; Capt. J. Bustanante, Spain; Capt. Manuel R. Nieto, Peru; Capt. T.P. Morgenstern, Poland; Commodore Olof F. Angelin, Sweden; Col. Karel Lukas, Czechoslovakia; Capt. E.C. Raguet, USN; Capt. R. Crenshaw, USN; Brig. Gen. Armando Revoredo, Peru; Commodore H.D. Clark, USN; Capt. Edson C. Sherwood, Canada; Capt. C. Scoufopoulos, Greece; Cmdr. J.H. Deprez, France; Capt. L.E. Hebbeck, Great Britain. Third row, Lt. Cmdr. Aristides Rojas, Venezuela; Lt. Cmdr. Felipe Cadenas, Cuba; Cmdr. Lin Sian-Kwan, China; Lt. Cmdr. P.R. Kittredge, USNR; Cmdr. H.E. Seidel, USN; Capt. H. Baltazzi, USN; Captain W.A. Hard, USN, Capt. Robert Henderson, USN; Capt. F. Ferris, USN; Capt. W.D. Wright, USN; Cmdr. Wilson Starbuck, USN, Lt. Cmdr. L.S. Tailor, USN; Cmdr. Donald McIntyre; Wing Commander Arturo Maneses, Chile; Lt. Cmdr. Daniel Candia, Paraguay; Lt. Jorge de Noronha, Brazil; Lt. James S. Copley, USNR.
Courtesy Sargeant Memorial Collection, Norfolk Public Library

BOTTOM LEFT: President Harry S. Truman and New York Mayor Fiorello La Guardia visit Norfolk Naval Operating Base, circa 1945.
Courtesy Sargeant Memorial Collection, Norfolk Public Library

ABOVE: The ship's company of the aircraft carrier USS Shangri-La (CV 38) musters on its flight deck in the Pacific. Built at the Norfolk Navy Yard and commissioned in 1944, the ship took part in strikes on Okinawa and Japan during the war and in atomic bomb testing afterward.
Courtesy Official US Navy Photo, Amy Waters Yarsinske Collection

RIGHT: Radio training class, 1945.
Courtesy Hampton Roads Naval Museum

ABOVE: Pilot Art Kreutz of Bombing Fighting Squadron 3 in front of his fighter plane at Oceana Naval Auxiliary Air Station. The squadron flew the F6F-5 Hellcat and was assigned to Carrier Air Group Three on USS Yorktown (CV 10) in the Pacific. On February 16, 1945, members became the first Navy carrier-based pilots to strike the Japanese homeland. They were awarded the Presidential Unit Citation. The squadron is the ancestor of today's Strike Fighter Squadron 32, the Swordsmen. Courtesy Norman Edge

TOP: Lt. j.g. Billy B. Chapman in a SB2C Helldiver, part of Torpedo Squadron 74 on the USS Midway (CVB 41), circa 1946. Courtesy Cynthia Chapman

ABOVE: Naval personnel disassemble a naval fighter plane for its parts, 1946. On the floor, from left, are R. M. Reynolds, aviation machinist's mate 2nd class; J. L. Yarbrough, aviation machinist's mate 2nd class; and P. M. Ford, aviation metalsmith 1st class. On wings, from left, are P. T. Smith and W. H. Covington, both seamen 1st class. Courtesy Sargeant Memorial Collection, Norfolk Public Library

BELOW: As Japan surrendered, Norfolk prepared to welcome home Army Col. John H. Lackey Jr., the first American aviator to land in Japan after the war's end. Courtesy Virginian-Pilot archives

Norfolk Virginian-Pilot

Vol. CLXXVIII, No. 63 — Norfolk, Virginia, Sunday, September 2, 1945 — Ten Cents the Copy

Japan Signs Surrender; Today Is Proclaimed V-J Day

Stony-Faced Aides of Emperor Write 'Unconditional' Defeat; Truman Hails Victory by Allies

Bowles Backs Budget Law For Jobless

Snyder Also Joins Plea Nation Needs Legislation To Stabilize Economy

MADE FIRST OFFICIAL LANDING—Col. John H. Lackey, of Norfolk ...

Congress Due Back at Work Wednesday

Truman to Drive for Action on Many Vital Reconversion Measures

Day of Retribution for Nipponese, World Victory, President Says

Day to Remember

We Shall Not Forget Pearl Harbor, Chief Executive Declares

Truman, M'Arthur and Nimitz Proclaim World Is at Peace

Many Famous Figures on Board Missouri for Historic Ceremony

12 Names Affixed

Marks First Setback for Nippon in Its 2,600-Year-Old History

RIGHT: Enjoying a little down time at the Navy YMCA, 1946. Courtesy Sargeant Memorial Collection, Norfolk Public Library

BOTTOM RIGHT: Wives and friends of crew members wave goodbye as Operation Highjump, the Navy's largest expedition to Antarctica, leaves Norfolk, December 2, 1946. Led by the flagship Mount Olympus, the expedition consisted of 13 ships and 4,000 sailors. It served as a cold-weather training exercise and photographed much of the Antarctic coast for the first time. Courtesy Virginian-Pilot archives

BELOW: Aerial view of Norfolk Naval Operating Base, 1946. Courtesy Official US Navy Photo, Virginian-Pilot archives

LEFT: Personnel inspection on the USS Coral Sea (CVB 43), the first postwar aircraft carrier, circa 1947. Courtesy Charles B. Lavender, USN, Retired

BOTTOM LEFT: AJ Trainer on the deck of the newly commissioned aircraft carrier USS Wright (CVL 49), 1947. Courtesy Charles B. Lavender, USN, Retired

BELOW: A torpedo bomber on the USS Wright, 1947. Courtesy Charles B. Lavender, USN, Retired

RIGHT: After the war, Hampton Roads remained a hub for naval aviation, including maintenance. The Overhaul and Repair Department at Norfolk Naval Air Station carried out much of that work. In this shop, workers examine, repair, overhaul and modify small detachable metal parts, such as stabilizers, wing flaps, rudders, ailerons, elevators and wing tips. They also remove defective rivets, apply corrosion inhibitor; and manufacture small items. 1947.
Courtesy Hampton Roads Naval Museum

BOTTOM RIGHT: This metal shop's job is to manufacture steel and aluminum aircraft parts, make the jigs and fixtures utilized for assembly, and perform bench work and welding. Courtesy Hampton Roads Naval Museum

ABOVE: Workers disassemble, clean and test the hydraulic equipment used with aircraft engines, such as pumps, valves and fuel tank pressure controls. Courtesy Hampton Roads Naval Museum

BELOW: Workers clean aircraft engines and their components using solvents, frit or vapor blasting, buffing and polishing. Courtesy Hampton Roads Naval Museum

ABOVE: Aircraft propellers are straightened, polished and balanced. Courtesy Hampton Roads Naval Museum

BELOW: In the fabric shop, workers repair fabric airfoils and make and repair a variety of other parts, from arm rests and map cases to safety belts and propeller covers. Courtesy Hampton Roads Naval Museum

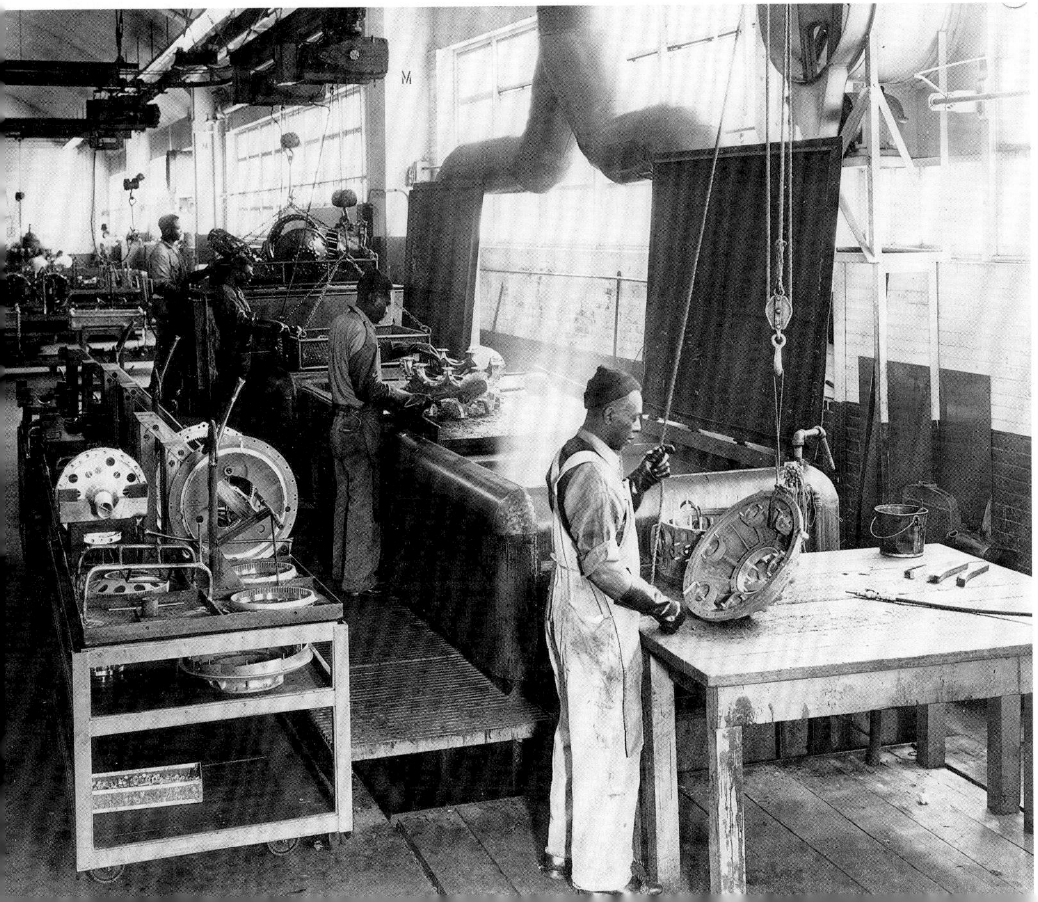

OPPOSITE: Airplanes are "canned" - placed in climate-controlled storage - at Norfolk Naval Air Station, 1948. Plans called for 328 planes to be placed in these special storage units so they could be pulled out and readied in a few hours when needed. Courtesy Sargeant Memorial Collection, Norfolk Public Library

BELOW: Lt. j.g. Falvey McKee Sandidge Jr.'s daughter Patty tries on her daddy's cover, 1947. Sandidge was a fighter pilot who saw action from World War II to Vietnam. Courtesy The Amy Waters Yarsinske Collection

RIGHT: A captured German V-2 rocket is fired from the flight deck of the aircraft carrier Midway (CVB 41) on September 6, 1947, marking the first time that a bombardment rocket had been launched from a ship or moving platform. The launching, held at sea off the East Coast of the United States, was observed by leading military and civilian personnel in the field of guided missiles. The rocket traveled about six miles and exploded. Courtesy Official US Navy Photo, Virginian-Pilot archives

ABOVE: An October 1947 Navy Days celebration, featuring public tours of some 50 vessels, ranging in size from the aircraft carrier Midway to small but speedy patrol craft. Courtesy Virginian-Pilot archives

MAX. CAP.
20 BOOM
11 RAD.

ABOVE: The aircraft carrier USS Franklin D. Roosevelt (CVB 42) leaves Norfolk for a Mediterranean deployment, 1948. Two years earlier, on July 21, 1946, off the Virginia Capes, the carrier made history when it launched the first jet from a carrier deck. The McDonnell XFD-1 Phantom jet was flown by Lt. Cmdr. James J. Davidson The aircraft, later renamed the FH-1 Phantom, was the Navy's first all-jet aircraft and the first to fly at 500 mph.
Courtesy Sargeant Memorial Collection, Norfolk Public Library

TOP RIGHT: The postwar period saw rapid developments in helicopter technology. This experimental helicopter XHJS-1, built by Sikorsky in 1948, was designed for utility work and incorporated several new features, such as all metal rotor blades, specially stressed landing gear for landing on the decks of ships and the elevation of the tail rotor.
Courtesy Official US Navy Photo, Virginian-Pilot archives

RIGHT: The USS Midway arrives home in Norfolk from the Mediterranean, 1948.
Courtesy Sargeant Memorial Collection, Norfolk Public Library

LEFT: The last of the Midway class of large carriers, the USS Coral Sea (CVB 43), leaves Norfolk a few months after its commissioning, 1948. It was named for the 1942 battle that marked the turning point of the war in the Pacific. Courtesy Virginian-Pilot archives

BOTTOM RIGHT: The R6V Constitution, one of two built for the Navy to serve as a long-range, high-capacity transport and passenger plane, 1949.
Courtesy Sargeant Memorial Collection, Norfolk Public Library

ABOVE: A P2V Neptune makes a jet-assisted takeoff from the aircraft carrier USS Midway (CVB 41), 1949. Courtesy Official US Navy Photo, Virginian-Pilot archives

CHAPTER THREE

The 1950s & '60s

The postwar peace came to an abrupt end in June 1950 with North Korea's invasion of South Korea. Just two days later, President Harry Truman ordered naval and air forces to the Far East. By September, the Norfolk-based aircraft carrier USS Leyte was headed to Korea.

One of the pilots aboard was Ens. Jesse L. Brown of Fighter Squadron 32, the Swordsmen. His fame as the first African American naval aviator was tragically cut short when he was shot down just a few months later. A fellow Swordsman, Lt. Thomas J. Hudner, went on to receive the Medal of Honor for his actions trying to save Brown.

Back home in Princess Anne County, the Navy was spending millions to transform the naval auxiliary air station at Oceana into the first master jet base on the East Coast – a decision that would shape Hampton Roads for decades to come.

In 1952, Norfolk welcomed a new aircraft to the fleet, the P5M-1 Marlin seaplane - a submarine "hunter-killer." In 1954, the carrier USS Hancock launched the first jets from a steam catapult, an innovation that allowed aircraft to carry heavier loads.

The Cuban Missile Crisis in the fall of 1962 brought nuclear tensions much closer to home. Adm. Robert L. Dennison, the commander in chief of U.S. Atlantic Command and the Atlantic Fleet, ordered his forces to blockade the island. Civilians at Guantanamo Bay Naval Base evacuated to Norfolk.

By the mid-'60s, Oceana-based squadrons were flying combat missions over Vietnam. In July 1967, the Norfolk-based carrier USS Forrestal had just arrived off Vietnam's coast when a fire broke out on the flight deck. More than 130 sailors died but the crew managed to save the ship through "far more acts of sheer heroism than I could count," the skipper said.

OPPOSITE: The USS Enterprise (CVN 65) shortly after it was christened in 1961. Courtesy Virginian-Pilot archives

RIGHT: Aircraft assembly area, March 1950.
Courtesy Hampton Roads Naval Museum

BOTTOM RIGHT: Planes being loaded aboard the USS Midway (CVB 41) at Norfolk Naval Base in preparation for an upcoming Mediterranean cruise, 1950.
Courtesy Sargeant Memorial Collection, Norfolk Public Library

BELOW: Ens. Jesse L. Brown, the first African American to be trained by the Navy as an aviator, was flying a Vought F4U-4 Corsair from the deck of the USS Leyte (CV 32) on a strike against the Chosin Reservoir in Korea on December 4, 1950, when he was shot down at low altitude. North Korean troops advanced on his downed Corsair, with Brown helplessly pinned inside. His wingman, Lt. j.g. Thomas Hudner, landed his Corsair wheels-up near Brown in a desperate attempt to free him; he was unsuccessful. Brown died of his wounds in the aircraft. He was later awarded the Distinguished Flying Cross.
Courtesy Official US Navy Photo, Amy Waters Yarsinske Collection

Norfolk Virginian-Pilot

Vol. CVC. No. 86 — Norfolk, Virginia, Sunday, June 25, 1950 — Price Fifteen Cents

Reds Invade South Korea

Hope Fades For 58 On Plane Feared Crashed In Lake Michigan

Smith Beats Graham in Senate Race

Truman's Foe Easy Winner In Big Upset

Victor Piles Up Huge Lead to Grab Nomination of Democrats

Hard, Bitter Fight

Search Fails To Find Trace Of Airliner

Capacity Filled Craft Is Believed to Have Crashed Into Storm-Swept Waves

Wreckage Sighted

Undeclared War Breaks Over Border

11 Points Pierced on Wide Front; Beaches Also Stormed

But No Air Bombs

Four Rail Lines Order Trains Be Cancelled

Flying Disc Saved

Pretty Fiancee, Pastor Aboard Missing Plane

Soviet Backs Invasion, Korean Diplomat States

Census Shows Suburb Trend

Today's Chuckle

Korea, Hot Spot In Cold War, Is Now in Battle

French Cabinet's Collapse Threatens Schuman Plan

Northeast North Carolina Gives Smith Big Majority

French Veterans Ask Big Pensions

ABOVE: A veteran of Admiral Halsey's Third Fleet in World War II, the aircraft carrier Wasp later served as an anti-submarine warfare support carrier. Courtesy Official US Navy Photo, Virginian-Pilot archives

LEFT: Early news reports didn't know quite what to make of the Communist attacks on South Korea. U.S. military advisors "expressed the opinion…that this might be the long-feared all-out invasion effort." Courtesy Virginian-Pilot archives

BELOW: Crowds gather to welcome home the USS Midway, 1950. Courtesy Virginian-Pilot archives

ABOVE: Navy planes, fully loaded with bombs, bullets and napalm, launch for strikes on Communist troops in Korea, 1951. Courtesy Hampton Roads Naval Museum

BELOW: Capt. Joseph Quilter, left, the commanding officer of Oceana Naval Auxiliary Air Station, celebrates the base's re-opening, September 19, 1951. New runways and other improvements helped transform Oceana into the Navy's first East Coast master jet base. Courtesy Official US Navy Photo, NAS Oceana archives

ABOVE: Lt. j.g. Thomas J. Hudner, who tried to rescue fellow pilot Ens. Jesse Brown during an attack on Chosin Reservoir, shakes hands with President Harry S. Truman after being presented with the Medal of Honor, April 13, 1951. Courtesy Hampton Roads Naval Museum

BELOW: L.W. Westcott, the first jet pilot to land after the re-opening of Oceana, September 20, 1951. The Navy spent about $12 million on the upgrades. Courtesy Official US Navy Photo, NAS Oceana archives

ABOVE: Midshipmen arrive at Norfolk Naval Air Station aboard an R6V Constitution, 1951.
Courtesy Sargeant Memorial Collection, Norfolk Public Library

LEFT: From left: Lt. Cmdr. Harold E. Rice, skipper of the submarine USS Diablo (SS 479); Lt. j.g. Kermit R. Sutliff; and Airman Alvin Baron aboard the sub docked at Norfolk Naval Base, September 6, 1951. The two airmen crashed their torpedo bomber 80 miles off Cape Henry in the Atlantic Ocean. The sub rescued them but was unable to find their third crew member.
Courtesy Sargeant Memorial Collection, Norfolk Public Library

ABOVE: The carrier Franklin D. Roosevelt (CV 42) is welcomed home from the Mediterranean, 1952. Courtesy Virginian-Pilot archives

RIGHT: Mrs. Francis J. Grisko, wife of the commanding officer of Patrol Squadron 44, prepares to christen a P5M-1 Marlin seaplane in April 1952. The aircraft was built for anti-submarine warfare. With her, from left: Rear Adm. Grover B. H. Hall, Cmdr. Grisko, John F. Floberg, Assistant Secretary of the Navy for Air, and Vice Adm. John J. Ballantine. Courtesy Ocean View Station Museum

ABOVE: The aircraft carrier USS Lake Champlain (CVS 39) catapults a pair of F2H-2 Banshee jet fighters for her first Korean War strikes, June 1953. Courtesy Hampton Roads Naval Museum

ABOVE RIGHT: The carrier USS Hancock (CVA 19) launches its first jet fighter, an F2H-3 Banshee piloted by Lt. William T. Brooks, using the Navy's new steam catapult in June 1954. The catapult enabled the Navy to launch aircraft carrying much heavier loads than could be handled on hydraulic catapults. Courtesy Albert E. Grimstead

BELOW: A P5M-1 Marlin seaplane swings at anchor while another anti-submarine weapon, a Navy blimp, passes overhead. The first P5M-1 was delivered to the fleet at Norfolk in April 1952. It was the first airplane designed and built especially for submarine warfare. Courtesy Ocean View Station Museum

ABOVE: Lt. W. P. Sullivan, left, and Capt. B.C. McCaffree prepare to cut the cake celebrating the 1,000th landing on the carrier USS Wasp (CV 18), January 11, 1952.
Courtesy Virginian-Pilot archives

OPPOSITE: Oceana aviators from Fighter Squadron 21 celebrate on April 2, 1954, after setting a new cross-country speed record in their F9F-7 Cougars. The three pilots flew from San Diego to New York in just over 3 hours and 45 minutes. Pictured left to right: Mrs. Francis X. Brady; Lt. Cmdr. Francis X. Brady; Rear Adm. Frederick N. Kivette, Lt. j.g. John C. Barrow; and Lt. Wallace "Buddy" Rich. The children are Mary Anne, James and John Brady. Courtesy Official US Navy Photo, Amy Waters Yarsinske Collection

ABOVE: Dental services at the dental office (Building Q-4) Oceana Naval Air Station, 1953. Courtesy Official US Navy Photo, NAS Oceana archives

TOP RIGHT: Adm. John H. Towers dedicates Ely Hall at Norfolk Naval Air Station, January 18, 1954. The building honors the memory of Eugene B. Ely, the first aviator to land and take off from a ship. Courtesy Dick Hanna

RIGHT: The Korean armistice in July 1953 meant an end to the war but President Eisenhower warned Americans not to let their guard down. Courtesy Virginian-Pilot archives

Weather Forecast

Norfolk Virginian-Pilot

Tides

Vol. CCVIII. No. 27 | Telephone 51431 | Norfolk, Virginia, Monday, July 27, 1953 | Price Five Cents

Allies, Reds Sign Korean Armistice

Cuba Quickly Ends Revolt; Many Killed

Men Wearing Military Uniforms Fail in Effort To Seize Barracks

Politicians Jailed

Exiled Ex-President Blamed by Batista In Uprising

Various Stages of Korean Conflict

Don't Relax, Eisenhower Tells Nation

Advises America To Be 'Vigilant Against Untoward' Events

'Solemn Gratitude'

'Looking Toward Unification of Korea;' President Quites Lincoln

By Douglas B. Cornell

U. S. Delegate Signs Korean Truce Document

All Shooting Scheduled to Stop Today

Rhee's Government Is Not Represented at Historic Ceremony

18 Copies of Truce

Exchange of Prisoners Expected to Start Within Week

By Robert B. Tuckman

ABOVE: Naval airmen in the control tower manage the evacuation of warplanes at Oceana Naval Air Station due to the approach of Hurricane Diane, August 1955. Courtesy Sargeant Memorial Collection, Norfolk Public Library

BELOW: The 60,000-ton supercarrier Forrestal (CVA 59) (top) passes her sister ship, Saratoga (CVA 60), in Hampton Roads, August 1956. The Forrestal was headed to sea for post-overhaul exercises. In the background is a portion of Norfolk Naval Station being dredged to support three piers for the supercarrier fleet. Courtesy Virginian-Pilot archives

ABOVE: Navy officials dedicate Apollo Soucek Field, named for the late vice admiral and head of the Navy's bureau of aeronautics, at Oceana Naval Air Station, June 4, 1957. Courtesy Official US Navy Photo, NAS Oceana archives

LEFT: Crew members celebrate the 10th anniversary of a world record-setting flight by the "Truculent Turtle," a P2V Neptune patrol aircraft that flew from Perth, Australia, to Columbus, Ohio, in 55 hours and 17 minutes in September 1946. Pictured are Capt. W.S. Reid, Capt. E.P. Rankin and Cmdr. R. H. Tabeling. Courtesy Dick Hanna

ABOVE: Aerial view of VIII Housing at Oceana, 1957. Courtesy Official US Navy Photo, NAS Oceana archives

ABOVE: The 60,000-ton supercarrier USS Ranger (CVA 61), the third of the Forrestal class, prepares for her sea trials and commissioning, 1957. Courtesy Virginian-Pilot archives

BELOW: Escort ships at sunset in the Mediterranean Sea, viewed from the flight deck of the USS Forrestal (CVA 59), 1958. Courtesy Thomas Scott

ABOVE: A big banner declaring 'We Love You All' is draped from the USS Randolph (CV 15) as it returned from to Norfolk from the Mediterranean, February 1958. It appeared the feeling was mutual among the families waiting on the pier. Courtesy Virginian-Pilot archives

ABOVE: Lt. J.R. Hawvermale, an F8U Crusader pilot at Oceana Naval Air Station, and Mrs. Joy Medlin make an appeal for government bonds, July 1958. Courtesy Virginian-Pilot archives

TOP LEFT: The flight deck of the aircraft carrier USS Antietam (CVA 36) bristles with helicopters and other aircraft as loading was completed at Norfolk Naval Station, 1958. The carrier USS Forrestal can be seen in the distance. Courtesy Virginian-Pilot archives

LEFT: A crowd waits to welcome the USS Randolph (CV 15) as it returns home in 1958. Courtesy Virginian-Pilot archives

Weather Today
Sunny – Cold
A.M. 25, P.M. 36
Thursday A.M., 20
Winds N. 15-20
Details and Tides on Back Page

The Virginian-Pilot

Largest Metropolitan Circulation in the Virginias and the Carolinas

Vol. CCXXIX. No. 71 — Telephone (Norfolk) MAdison 5-1411 — Norfolk, Portsmouth, South Norfolk, Virginia, Wednesday, December 10, 1958 — 38 Pages — Price Five Cents

2 Infants, 4 Airmen Die in Crash on Granby

Tragedy Mars Record For Civilians' Safety

• Two Homes, Truck Destroyed as Plane Falls Short of Air Station Runway

By Richard M. Mansfield, Virginian-Pilot Military Reporter

NORFOLK — A crippled Navy aerial tanker plunged into two North Granby street homes Tuesday snuffing out the lives of four naval airmen and two infants. It ended a miraculous history of non-civilian deaths from military air crashes in Norfolk.

Loaded with jet fuel and aviation gasoline the plane, an AJ-2 twin-engine Savage, had experienced engine trouble during a refueling exercise. It was returning to the Norfolk Naval Air Station when the crash occurred in the 8900 block of Granby street.

The pilot was on a right turn on final approach to Runway 10 when his left engine failed, at about 12:14 p.m.

According to eye-witnesses the plane's right wing swung down to a "90 degree bank," and appeared to "be in trouble" just before it struck. The right wing hit the edge of the roof of the house at 8952 Granby, sliced through the body of a potato-laden tractor-trailer driving north on Granby and catapulted into the two houses at 8939 and 8943 Granby.

Mother Rescued

In the second house was Mrs. Joseph E. Tondora and her infants, 14-day-old Michalene and 14-month-old Joseph Jr. Mrs. Tondora was rescued but the babies were killed. She was taken from the flaming house by three men who rushed onto the scene. They restrained her from returning to the house.

Two of the men, one identified as N. E. Jones of the Chesapeake and Potomac Telephone Co., rushed into the almost-demolished and burning house for the children but were forced out by smoke and flame. The third man, William W. Mason, 8412 Larrymore Ave., remained with Mrs. Tondora until she was taken to the NAS Dispensary.

The house at 8939 Granby was completely demolished, and unoccupied at the time of the tragedy. This is the home of W. C. Poore, who was scheduled to be released from DePaul Hospital at 11 a.m. Tuesday, but his release was delayed.

The four members of the AJ-2 crew were killed in the crash. They were all assigned to Heavy Attack Squadron 15 at NAS. The Navy identified them as:

Lt. Cmdr. George Wilson, 38, of 840 Carrene Dr., Birchwood Gardens. He is survived by his wife, two sons and a daughter. He was a native of Pacoima, Calif.

Ens. Fred M. Clancy III, 20, of Kissimmee, Fla. He is survived by his parents, Mr. and Mrs. Fred M. Clancy Jr., of the Florida address.

Crewman Patrick F. Toomey, 19, son of Mr. and Mrs. Francis J. Toomey of 1814 Plymouth Rd., Durham, N.C.

John E. Delaney, ADN3, 20, son of Mr. and Mrs. Timothy J. Delaney, 116 Clark Ave., Chelsea, Mass.

The tragedy marked the first time in history that civilians have been killed by military air crashes in the Norfolk area. Only last Jan. 16 a Navy P4M Mercator patrol plane crashed into several cottages in East Ocean View, claiming the lives of four crew members and injuring eight other crewmembers and three civilians.

For further reading see Crash, Page 17

Crew of Ill-Fated Plane

Wilson — Clancy — Toomey — Delaney

Tail of Plane in Scene of Death and Destruction on Busy Granby Street

The tail of the AJ-2 Savage aerial tanker, almost unscathed by the crash, stands by the inferno that claimed six lives Tuesday when the crippled plane plunged onto Granby street.

8,694-Mile ICBM Claimed by Reds

• Distance Would Surpass Best U.S. Missile, Put Any Spot in America Within Range

WASHINGTON (AP)—Soviet Premier Nikita Khrushchev was reported Tuesday to have told Sen. Hubert Humphrey (D-Minn) that Russia has an intercontinental ballistic missile capable of firing a warhead 8,694 miles.

Achievement of such a range would surpass the best U. S. ICBM effort so far—the Atlas which the Air Force hurled 6,325 miles Nov. 28. It also would put almost all the United States within reach of Soviet weapons.

Senate sources told of Khrushchev's "secret" message to Humphrey. The Minnesota senator himself declined to give details of his 30-minute briefing of President Eisenhower on his long talk with Khrushchev.

On Nuclear Blast

Humphrey told reporters he passed on to Eisenhower a confidential message from the Soviet Premier dealing with Russian nuclear explosions of a very substantial size. He also told Eisenhower about secret Soviet developments in the missile field, the senator said.

He said the administration would have to decide whether to make public the information he relayed from the Soviet Premier to Eisenhower. The White House said it would not do so if it contained too much Soviet propaganda.

White House press secretary James C. Hagerty stuck by that stand after Senate sources reported what Humphrey told Eisenhower about missiles. Hagerty said the government would have to evaluate the information before deciding.

The Pentagon had no comment

Labor Party In Offing For U.S.?

• Meany Says Unions Prepared To Be as Political as Necessary To Win Objectives

NEW YORK (AP) — AFL-CIO President George Meany told a state union convention Tuesday that organized labor is quite prepared to form its own political party if forced into it.

"Labor is going to be just as political as it has to be to win its objectives," he said in reference to current AFL-CIO political action programs.

"If labor has to go further in order to make America a better place in which to work and live, to make America the leader of the democracies of the world, we will take the next step if it is forced upon us.

Good Job

"I have always said we don't want our own national party, but if we have to do that to lick the people who want to drag us back

British Deny Division with West Germany

LONDON (AP) — British diplomats Tuesday denied reports of a rift between Britain and Germany over their Berlin policies.

They scoffed at a suggestion that West German Chancellor Adenauer had planned to fly to Britain this week to try to heal the breach.

Informants emphasized there is no divergence between Britain and Germany over their attitude toward Soviet Premier Nikita Khrushchev's free city proposals for isolated Berlin. Both nations turned them down.

Adenauer had planned to fly here Friday for talks with Prime Minister Macmillan on European trade problems but reports from Bonn Tuesday said his doctors ordered him to bed because of a feverish cold and the 82-year-old Chancellor called off his trip.

"There had been speculation that a second reason might be a split in views on Berlin. Adenauer was said to oppose Britain's view that the question of status should be included in future talks with the Soviets on Germany and Europe.

'Something to Think About'

Stories of Fright, Heroics on Granby

NORFOLK — Except for some fire and foam on one side, the T. H. Hill house at 8935 Granby, next door to the demolished homes, was virtually undamaged by the Navy plane crash Tuesday.

Mrs. Edna Bellamy, Negro, of 412 Charlotte St., Mrs. Hill's maid, was alone at the time of the crash.

On her way to the kitchen to see about something in the stove, she heard "a blast." "It knocked me down," she said, "and when I was getting up there were two more and I was knocked down again."

Followed by the Hill dog, Tag, a 14-year-old cocker spaniel, Mrs. Bellamy ran to the front door. "The whole front yard was nothing but fire," she recalled.

She then went to the back door which was open. But once again she found it impossible to move because of the flames. She went back inside until "two Navy men" carried her out.

Mrs. Hill was told later that the maid had picked up the dog and after they had been taken out, Mrs. Bellamy took her coat off and put it over the shivering animal.

Mrs. Hill and her daughter Betty, 14, were returning from Ward Corner and had reached Bay Ave., when they saw the fire. "I could tell it was near our house," Mrs. Hill said. "I tried to make Betty stay in the car, but she ran down the street with me after her. I don't even know where the car is."

Betty, a ninth-grade student at Oscar Smith at night, repeated over and over, "It's a miracle."

The Donald L. Foxwells, 8952 Granby, were not at home when the accident occurred. The plane's wing cut a gaping hole in the eaves of the house, knocking out all the wiring for the electrical system, before the aircraft plowed into the two houses across the street.

For further reading see Stories, Page 18

Other pictures, stories on pages 16, 17 and 19.

Illness Protects House Resident

• Occupant of Demolished Building, Truck Driver Have Narrow Escapes

By Betty Bradley

NORFOLK — A conscientious doctor and a few seconds of time saved the lives of at least two Tuesday.

W. C. Poore, a resident of the demolished house at 8939 Granby St., and Charles Sheppard Jr., driver of the potato-laden truck, narrowly missed being victims of the crash.

Poore was hospitalized Monday. Because he developed a cold, an operation set for today was postponed.

According to his physician, Poore would normally have returned to his home after checking out at 11 a.m. He would then have been in his house at 12:14 p.m. when the plane crashed.

However, the doctors put off Poore's discharge until about 1:45 p.m. in order to obtain "a history and a physical." Poore was told of the tragedy prior to leaving the hospital. His wife and daughter, Becky, 18, were working and his son, Woody Jr., 14, was in school.

Sheppard, 21, Negro, of Elephants Fork Rd., Suffolk, was on his way to Newport News, "just riding along like always," when something hit his truck "all of a sudden."

The right wing of the plane had torn into the potato-filled trailer, knocking Sheppard to the top of his cab and setting fire to both trailer and cab. If he had been going a fraction of a second slower, the wing would have struck the cab.

Sheppard tried to get out of one side of the cab but was blocked by the flames. Blinded by smoke, he was unable to see where the truck was going. It eventually stopped on a bank.

Although he had heard the plane, Sheppard had taken no notice of it. Only after he held his nose with his hand and groped his way through the fire-swept door did he know what had happened.

For further reading see Escape, Page 18

On Inside Pages

ABOVE: Workman haul away the tail of the Navy AJ-2 Savage tanker that careened into two homes in the 8900 block of Granby Street in Norfolk on December 9, 1958. Six people were killed — the four crewmen and two babies in the home to the right of the plane tail. The two-engine craft was approaching the runway at Norfolk Naval Air Station for an emergency landing when its left engine failed. Its right wing hit the edge of a roof across the street, then a tractor-trailer, and the plane skidded into the homes, which burst into flames. The mother of the two babies was saved. Courtesy Virginian-Pilot archives

LEFT: Witnesses told The Pilot that the plane appeared to be in trouble just before it swung down, striking the roof of one house, slicing through the truck and colliding with two more homes. Courtesy Virginian-Pilot archives

BELOW: Potatoes spew from the tractor-trailer truck hit by the Navy tanker. The truck driver was treated for burns and released. Courtesy Virginian-Pilot archives

ABOVE: The amphibious assault ship USS Boxer (LPH 4), 1959. It was designed to transport and land helicopters, equipment and personnel.
Courtesy Official US Navy Photo, Virginian-Pilot archives

TOP LEFT: A large crowd turns out to welcome home the USS Intrepid (CV 11) as it pulls into port in Norfolk, 1955. Courtesy Virginian-Pilot archives

BOTTOM LEFT: An F4D Skyray jet fighter catapults off the flight deck of the USS Intrepid (CV 11), May 4, 1959. Courtesy Official US Navy Photo, Virginian-Pilot archives -

BELOW: A WF-2 Tracer, an airborne early warning aircraft, revs up for a launch from the USS Bennington (CV 20), 1959. Courtesy Official US Navy Photo, Amy Waters Yarsinske Collection

ABOVE: Moving planes into position on the aircraft carrier USS Valley Forge (CVS 45), the flagship of an anti-submarine task force that developed tactics to counter fast, deep-diving nuclear subs, 1958. Courtesy Official US Navy Photo, Virginian-Pilot archives

TOP RIGHT: A-4D Skyhawks in formation above the Mediterranean Sea, 1960. Courtesy Official US Navy Photo, Virginian-Pilot archives

RIGHT: The aircraft carrier USS Intrepid maneuvers along the Elizabeth River, headed for the Norfolk Naval Shipyard, 1960. Courtesy Virginian-Pilot archives

ABOVE: Girls from Virginia Beach High School on the Norfolk Naval Air Station entry in the Armed Forces Day parade in Virginia Beach, 1961. Courtesy Hampton Roads Naval Museum

BELOW: The Foreign Object Damage (FOD) cleanup crew begins its weekly tour of the base, 1962. Courtesy Hampton Roads Naval Museum

ABOVE: Construction work continues on the USS Enterprise, 1960. Courtesy Virginian-Pilot archives

BELOW: Armed Forces Day parade in Norfolk, 1961. Courtesy Hampton Roads Naval Museum

Fair and Cold Weather Due Throughout Area

NORFOLK **LEDGER-STAR** PORTSMOUTH

HOME EDITION

87th Year, No. 70 Five Cents Tuesday, October 23, 1962 Norfolk, Portsmouth, Va.

Warships Clamp Blockade on Cuba

Step Toward Nuclear War, Says Soviet

Kennedy's Charges Denounced

U.S. Fleet Fans Out Around Cuba

Showdown Coming Fast

By John M. Hightower

Cuban Crisis At-a-Glance

Hail, Stop, Search—and, If Necessary, Sink

Cuba Calls All Forces To the Colors

Stock Mart Trade Heavy

Guantanamo Emptied of Dependents

Norfolk Haven for Evacuees From Cuba

Around-the-Clock Operation

Navy Base Berths Emptied

Rusk Asks OAS To Back Quarantine

Other Cuba News And Viewpoints

Dover Air Base Cargo Planes Out

Hands of War Raised in Compassion

ABOVE: As Navy ships moved to blockade Cuba, hundreds of women and children fled Guantanamo Naval Base by air and nearly 2,000 more evacuees headed to Little Creek Naval Amphibious Base aboard ships. Courtesy Virginian-Pilot archives

ABOVE RIGHT: The USS Essex (CV 9) was among the carriers participating in the Cuban blockade. Courtesy Official US Navy Photo, Virginian-Pilot archives

RIGHT: Navy greeters eye a Boeing L1011 as it rolls up to a Norfolk Naval Air Station hangar carrying evacuees from Guantanamo Bay Naval Base Courtesy Virginian-Pilot archives

Norfolk Haven for Civilians Evacuated From Guantanamo Because of Blockade

Five plane loads of civilians arrived in Norfolk Monday night with 338 persons evacuated from the Guantanamo Naval Base in Cuba. The first climbed from a plane at the Naval Air Station shortly after 6 p.m. (left photo). Babes in arms were a common sight. Mrs. Georgia Bernoccio, of Asbury Park, N.J., (center photo) cuddles her infant after the group was whisked from the air station to temporary quarters at the Little Creek Amphibious Base. Others lounged in the lobby of Barracks 3602 awaiting accommodations for the night (right photo) and phone calls from relatives.

LEDGER-STAR

TUESDAY AFTERNOON, OCTOBER 23, 1962

Carl Cahill's Corner

Old Road To Work On Tedium Drive

Things Happened in a Hurry

Jerry and Susanne Work Talk Over Latest Cold War Turn

Gitmo People Moved Light, Fast in Crisis

By JIM STIFF

Flashlight His Concern

Michael Bixby Walks Patiently

Norfolk Offers Full Services As Officials Back Kennedy

Sign of Police

Girl Lost Again

Re-Apportionment Invalidation Plaintiffs' Aim

Norfolk Citizens' Votes Devaluated, Howell Argues in Redistricting Case

By TOM REILLY

Frost Expected Tonight As Cool Snap Persists

NORFOLK LEDGER-STAR PORTSMOUTH

LATE HOME EDITION

87th Year, No. 71 Five Cents Wednesday, October 24, 1962 Norfolk, Portsmouth, Va.

Soviet Ships Near U.S. Blockade

Russian Ships Photographed by United States, Which Reported the Vessels En Route to Cuba Carrying Jet Light Bombers in Crates

Almond Recess Appointee

Dennison Calling Shots

Norfolk Base Poised For Cuba Challenge

U.S. Gains Support in UN Council

Danger Warning Broadcast by Navy

By Elton C. Fay and Fred S. Hoffman

Cuba at a Glance

By THE ASSOCIATED PRESS

Help in Blockade Offered to U.S.

Soviet Returns Proclamation

Reds Renew Demand on West Berlin

In Today's LEDGER-STAR

Moscow Far Distant

U.S. Might, Reds' Long Haul Make Blockade Task Easier

By HANSON W. BALDWIN

TRANSPORTER IN HOLD REVETMENT

EMPTY TRANSPORTER

MISSILE ON LAUNCHER

MISSILE HOLD REVETMENTS

LAUNCHER

Defense Department Photo of Missile Base in Cuba

ABOVE & LEFT: Evacuees at Little Creek described leaving most of their possessions behind and having little idea where they were headed. "Some of you people selling papers should bring a stack of them out here and sell them quick so we'll know what's happening," said one.

Courtesy Virginian-Pilot archives

RIGHT: An RA-5C Vigilante from Reconnaissance Attack Squadron 5, the Savage Sons, is prepared for launch aboard the aircraft carrier USS America (CVA 66) in the Mediterranean Sea, 1965. The Vigilante was the Navy's first Mach 2, nuclear capable, carrier-launched bomber.
Courtesy Official US Navy Photo, Amy Waters Yarsinske Collection

BOTTOM RIGHT: Nicknamed the "Bantam Bomber," the A4D Skyhawk was the Navy's smallest atomic-weapon-carrying plane in the 1960s. The light bomber was manufactured by the Douglas Aircraft Corp. and was capable of taking off and landing aboard an aircraft carrier. Courtesy Official US Navy Photo, Virginian-Pilot archives

BELOW: A message scrawled on a bomb aboard the USS Independence (CVA 62), the first Atlantic Fleet carrier to conduct combat operations in Vietnam, 1965. Courtesy Official US Navy Photo, Virginian-Pilot archives

Viet Nam

Brezhnev Threatens Reprisal

Hanging Abolition Voted

Lords Back 5-Year Ban

2 Captured Pilots From Va. Beach

Reds Assert Pilot Downed

The Virginian-Pilot

Fighting Goes On As Marines Land

Viet Cong Stay Clear of Beach

Ala. Troopers Block March

WAR ESCALATES

Cubans Attack Soviet Embassy in D.C.

Guards Caught by Surprise

Bonn Acts To Heal Israel Rift

Board President Says:

Central Y to Accept All Races

Descendant of Victoria

Wife of Sweden's King Dies

SCIENTIST GOES DEEP INTO SLEEP

ABOVE: In March 1965, the Vietnam conflict intensified with the landing of U.S. Marines at Da Nang Bay. The number of naval aviators – some with ties to Virginia Beach - being shot down and taken prisoner would also grow. Courtesy Virginian-Pilot archives

LEFT: The aircraft carrier USS John F. Kennedy (CV67) during construction in Newport News, 1960s. Courtesy Virginian-Pilot archives

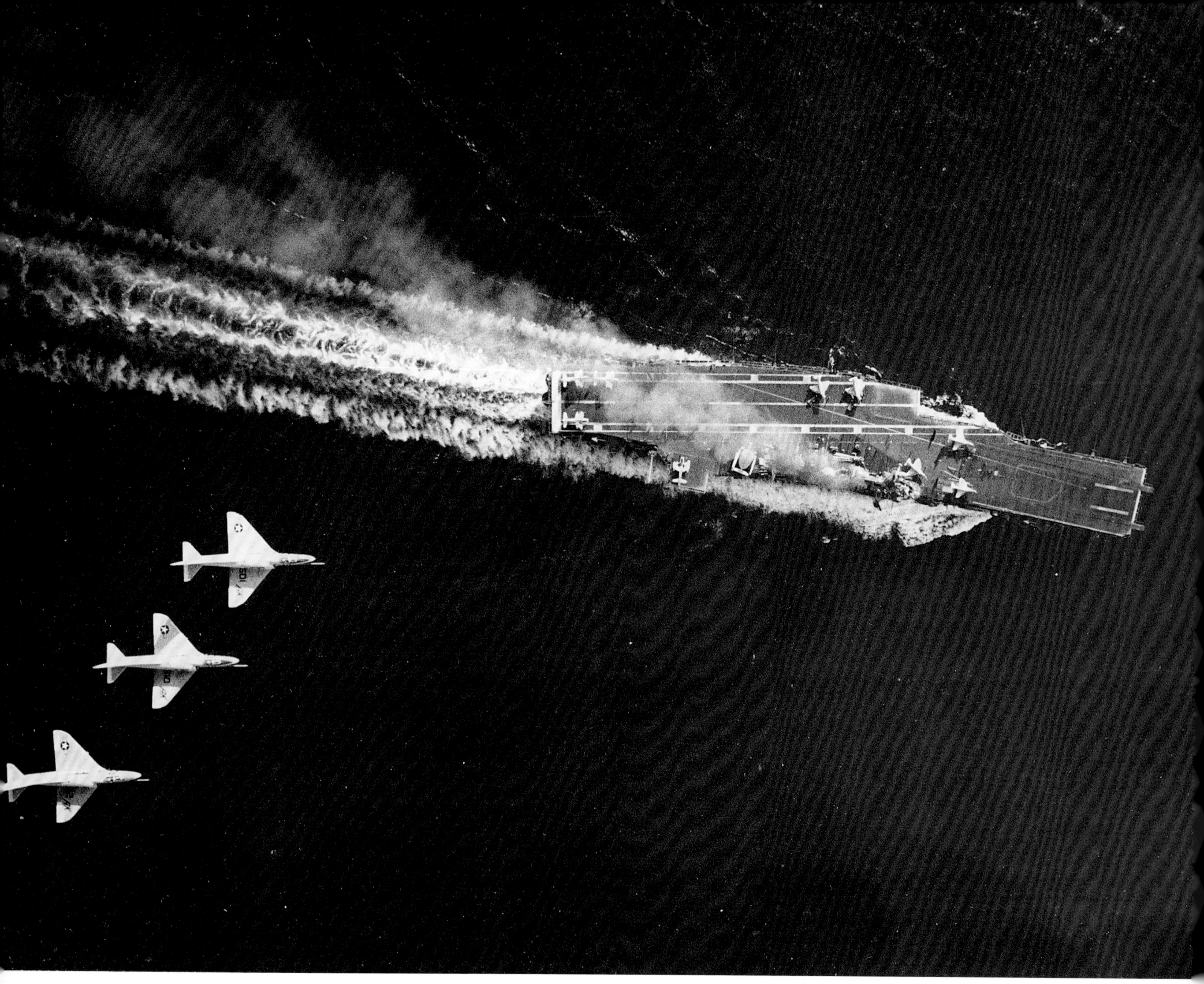

LEFT: Lt. John M. Waples, an A-4C Skyhawk pilot with Attack Squadron 76,the Spirits, returns to Oceana Naval Air Station from Vietnam, 1966. Virginia Beach Mayor Frank Dusch climbs the ladder to the cockpit with a glass of champagne for Waples. The skipper of the squadron, Cmdr. J. B. Linder, was awarded the Silver Star for leading a strike on a missile site at Thanh Hoa. Courtesy Official US Navy Photo, Amy Waters Yarsinske Collection

BOTTOM LEFT: Millie Earnest (left), wife of incoming Oceana commanding officer Capt. Albert K. Earnest, shares a happy moment with Helen Rumsey (center), spouse of the outgoing commander, Capt. Dexter C. Rumsey II. Rumsey is holding one of the many gifts given to her at the hail-and-farewell event traditionally accorded incoming and outgoing commanding officers and their spouses. At right are Mrs. L. H. Petersen and Mrs. P. N. Moriarty. Courtesy The Amy Waters Yarsinske Collection

BOTTOM RIGHT: Lt. Ronald F. Bell is rescued from the Tonkin Gulf after ejecting from his disabled F-8 Crusader. Courtesy Patricia Bowler

OPPOSITE: In formation over the USS Intrepid (CVS 11), three A-4 Skyhawks from Attack Squadron 95, the Green Lizards, return from combat assignments over North Vietnam, 1966. Courtesy Official US Navy Photo, Virginian-Pilot archives

RIGHT: Nearly four years after the assassination of President John F. Kennedy, Defense Secretary Robert S. McNamara, and Kennedy's brothers Robert and Edward arrive for the christening of the aircraft carrier named in his honor, May 27, 1967. Courtesy Virginian-Pilot archives

BOTTOM RIGHT: The ceremony took place at Newport News Shipbuilding, where the carrier was built. Courtesy Virginian-Pilot archives

OPPOSITE: The President's 9-year-old daughter, Caroline, smashes a bottle of champagne against bow of the ship as her mother, Jacqueline, and brother, John, look on. President Lyndon Johnson is at far left; shipyard President Donald A. Holden is behind Caroline. Courtesy Virginian-Pilot archives

BELOW: Large crowds turned out to watch the ceremony. Courtesy Virginian-Pilot archives

RIGHT: An A-6A Intruder from Attack Squadron 35, the Blue Blasters, awaits its signal for launch on the USS Enterprise, 1967. Three steel-plated jet blase deflectors are hydraulically lifted behind the plane to prevent damage to other aircraft as the jets turn up to full power. Courtesy Virginian-Pilot archives

BELOW: The Intruder was a low-level, all-weather attack bomber designed to operate night and day, from land or sea. It was replaced by the F/A-18 Hornet. Courtesy Official US Navy Photo, Virginian-Pilot archives

Hot and Humid
High Today 90
Low Tonight 72
S-SW, 10-20 m.p.h.

The Virginian-Pilot

Today, Editorially
• Educating Educators
• Farm Handouts
• Keep the Lights On
Page C-4

102nd Year. No. 251 — Norfolk, Portsmouth, Virginia Beach, and Chesapeake, Virginia, Sunday, July 30, 1967 — 134 Pages Price 20 Cents

46 Killed, 56 Hurt In 'Forrestal' Fire

80 Also Feared Lost

More pictures and stories on Page A-3.

Crewmen fight fires in planes on flight deck. (AP)

Va. May Issue Bonds

By GEORGE M. KELLEY

FOCUS . . .

. . . ON THE WORLD

Saudi Arabia, Iraq Resume Oil Pumping

Nigerian Unity Tough, Costly

Maoists Regain Control in Wuhan

. . . ON THE NATION

Miami Beach Site for GOP Convention

Navigation System Allowed for Civilians

New Farm Standard Sought

Paratrooper Shoots Negro in Detroit

Johnson Tax Package Expected Soon

. . . ON THE METRO AREA

Dr. Reich to Be Rabbi in N.Y.

Vandals Rough on Construction Firm

. . . ON THE SPORTS WORLD

Matson, Nelson Pace U.S. Trackmen

'Liberty' Brings In Memories

Skipper Describes Attack

By CLIFFORD HUBBARD

Delighted Cmdr. McGonagle welcomes Lentini back aboard.

Get Truth On Riots, Panel Told

Today's Chuckle

Liechtenstein Wedding

Fairy Tale of 2 Titles Today

Liechtenstein's prince dances with Greece's queen. (AP)

ABOVE & TOP: Forrestal fire, July 29, 1967. Courtesy Virginian-Pilot archives

LEFT: On July 29, 1967, the aircraft carrier USS Forrestal was about to launch its second combat aircraft of the day when a rocket misfired and struck a fuel tank on the flight deck. Some crew members perished in the explosion; others were trapped beneath the flight deck. More than 130 sailors died. The disaster reshaped fire safety and firefighting training in the Navy. Courtesy Virginian-Pilot archives

DANGER
KEEP AWAY
NAVY
SH-3D
156501
HS3
001
USS FORRESTAL
6501

CHAPTER FOUR

The 1970s, '80s, & '90s

Naval aviation marked its 60th anniversary in 1971, with the United States embroiled in the Vietnam War. The conflict's final years placed great demands on Navy air power, including squadrons based at Oceana Naval Air Station in Virginia Beach, which by then had cemented itself as the East Coast's master jet base. Even after the United States declared an end to military involvement in Vietnam in 1973, the job for naval aviators wasn't done. Some POWs – including pilots – still had not come home, and in 1975, the Navy was called upon to help evacuate Vietnamese refugees.

Also in 1975, the Navy commissioned the first of a new class of aircraft carriers, the Nimitz-class supercarriers. By the end of the 1990s, the service had launched seven more, all of them built in Newport News: the Dwight D. Eisenhower, the Carl Vinson, the Theodore Roosevelt, the Abraham Lincoln, the George Washington, the John C. Stennis and the Harry S. Truman. This era also saw the introduction of several important aircraft, including the F-14 Tomcat, which came to Oceana in 1974, and the F/A-18 Hornet, which the Navy began testing a few years later.

By the 1980s, worldwide acts of terrorism were on the rise, and global conflict zones were shifting. Over the course of the next 20 years, naval aviators played a role in crises in Iran, Libya, Grenada, Lebanon, the Persian Gulf and the Balkans. During 1991's Operation Desert Storm, Navy ships and aircraft struck countless targets in Iraq.

1991 also brought the collapse of the Soviet Union. With the United States securely positioned as the world's only superpower, naval aviation was downsized and many squadrons and shore commands were decommissioned or consolidated.

Also during this period, naval aviators – namely those flying the E-2C Hawkeye – began to assist federal agencies working to curb the flow of drugs into the country.

The 1990s saw the Navy's last A-6 Intruder flight, carried out at Oceana by Lt. Ronald Lee "Rhino" Wise. Wise's career later came to a tragic end; he was killed in July 1998 when his Tomcat collided with another aircraft over the Mediterranean Sea.

In April 1993, the secretary of defense moved to drop many of the restrictions that until then had kept women on the sidelines of naval aviation. The following year, the Eisenhower became the first carrier to deploy with women permanently assigned to the crew.

By the end of the '90s, the United States' global military superiority remained unquestionable – due in no small part to naval aviation.

OPPOSITE: Starving flood victims in Tunisia crowd under the whirling rotors of a helicopter from the USS Forrestal as it delivers bread and enriched fruit drinks, 1973. Courtesy Official US Navy Photo, NAS Oceana archives

ABOVE: USS America (CV 66) prepares to leave on its final Vietnam deployment, 1972. Courtesy Virginian-Pilot archives

TOP RIGHT: Crowds greet returning prisoners-of-war at Oceana Naval Air Station. Many Vietnam POWs were naval aviators, including James Bond Stockdale, the highest-ranking naval officer held prisoner in North Vietnam. He spent more than seven years in captivity and served as a leader to fellow POWs. He was later awarded the Medal of Honor and promoted to vice admiral. Courtesy Official US Navy Photo, NAS Oceana archives

RIGHT: Capt. Jeremiah Denton, the first East Coast pilot taken prisoner in North Vietnam, was among those welcomed home in February 1973. He spent seven and a half years in captivity. Denton went on to become a rear admiral. Courtesy Virginian-Pilot archives

Cloudy
High Today 46
Low Tonight 30
NE, 5-15 m.p.h.

The Virginian-Pilot

108th Year. No. 84

Norfolk, Portsmouth, Virginia Beach, and Chesapeake, Virginia, Tuesday, February 13, 1973

Today, Editorially
• Godwin Runs Again
• Homecoming
• Score in Overtime
Page A18

Price 10 Cents

143 Ex-POWs Taste Joys of Freedom

Yen to Float
U.S. Cuts Dollar By 10%

Favorite Foods, Calls

Capt. Denton is welcomed by Adm. Noel Gayler at Clark. (UPI)

Lt. Gen. William Moore Jr. greets Navy Cmdr. Ray D. Vohden of Memphis, Tenn. (UPI)

4 Tidewater Families Aglow Again

Howell Seeks Debates
Support Lines Up For Godwin Race

By GEORGE M. KELLEY

Abbitt

Denton, Tschudy, Mulligan, Cherry Homeward Bound

By JOHN STEVENSON

RIGHT: President Gerald Ford speaks at the commissioning of the USS Nimitz (CVN 68), the first in a new class of nuclear-powered aircraft carriers, May 3, 1975. Courtesy Virginian-Pilot archives

BOTTOM RIGHT: Members of the Princess Anne County Virginia Militia Hussars, dressed in colonial military attire, present a Bicentennial commemorative certificate to Capt. W. D. "Bill" Knutson, commanding officer of Oceana Naval Air Station, December 1975. Courtesy Official US Navy Photo, NAS Oceana archives

BELOW: Another group of POWs, six with ties to Hampton Roads, was set to be released in March 1973. Courtesy Virginian-Pilot archives

Sunny, Mild
High Today 70
Low Tonight 46
NW, 10-20 m.p.h.
For current report call 836-1212

The Virginian-Pilot

108th Year. No. 112 — Norfolk, Portsmouth, Virginia Beach, and Chesapeake, Virginia, Tuesday, March 13, 1973 — Price 10 Cents

Today, Editorially
- Nixon 'Gets Tough'
- Monetary Muddle
- Battle for No. 2

Page A14

140 U.S. Captives To Get Freedom

Turnover Planned Wednesday, Friday

By GEORGE ESPER

SAIGON (AP) — The U.S. Command Monday received from North Vietnam the names of 108 American prisoners of war to be released Wednesday. The Vietcong announced that they will release 32 POWs Friday at Hanoi's Gia Lam Airport.

A spokesman for the Vietcong delegation said the list of names will be presented to the United States at a meeting of the four-party subcommission on prisoners today.

The Vietcong initially indicated that they would release their next bath of prisoners Wednesday, then changed the day to Thursday.

The spokesman said the list includes Maj. Theodore William Gosta, 34, of Sheridan, Wyo., who is suffering from a "nervous illness." He did not elaborate Gosta was captured Feb. 1, 1968, during the Communists' Tet offensive.

The North Vietnamese delegation handed the United States its list of prisoners Monday.

The Defense Department relayed the news to families across the United States, some of whom have waited up to six years through a long and bitter war.

When the Vietcong hand over their list of prisoners to the U.S. delegation, it, too, will be immediately transmitted to Washington for notification of the men's families.

The captives of North Vietnam will be turned over to U.S. authorities at Gia Lam and flown to Clark Air Base in the Philippines for the jubilant welcome previous groups of returning prisoners have enjoyed.

The two groups are among 298 American prisoners still held by the Communists. Their release, pegged to U.S. troop withdrawals, will complete about 75 per cent of the release of American prisoners.

The last group is to be freed by March 28, the deadline set by the Vietnam peace agreement for release of all American prisoners and for withdrawal of all U.S. troops.

The Communists so far have released 299 American prisoners —Feb. 12, Feb. 18, and March 5. When the Vietnam agrement was signed in Paris Jan. 27, the Communist delegations turned over to the United States lists of 585 American prisoners held in North Vietnam, South Vietnam, and Laos.

The United States said it anticipates no trouble in completing repatriation by the deadline. But it temporarily suspended withdrawal of the fewer than 7,000 American troops remaining in Vietnam as insurance that the prisoner release will keep pace with the withdrawal.

Maj. Gen. Gilbert H. Woodward, senior U.S. representative on the four-party Joint Military Commission (JMC), told the North Vietnamese and Vietcong delegations that there will be no further withdrawals until the American prisoners are turned over Wednesday and Friday.

Son on List
Ex-Pacific Boss Waits

Adm. McCain

Lt. Cmdr. McCain

By HUGH MULLIGAN

SAIGON (AP)—A Navy pilot scheduled to check out of the Hanoi Hilton this week is the son and namesake of the admiral who directed some of the heaviest bombing against North Vietnam.

Lt. Cmdr. John S. McCain III, 36, was shot down in his Skyhawk jet by a SAM missile over North Vietnam Oct. 26, 1967, eight months before his father, Adm. John S. McCain Jr., took over in Honolulu as commander of U.S. forces in the Pacific.

(Adm. McCain served in Norfolk from August 1960 through May 1961 as chief of Amphibious Group 2 at Little Creek. Then he took over the Amphibious Training Command, also at Little Creek, and remained until August 1962.

(The admiral's final tour in Tidewater was from September 1963 to June 1965 as commander of the Atlantic Fleet Amphibious Force.)

The pilot's grandfather, Vice Adm. John Sid-

(See McCains, Page A7)

6 of Tidewater To Be Released

By JOHN STEVENSON
Virginian-Pilot Staff Writer

NORFOLK—Six more men with Tidewater ties are on the list of American prisoners of war to be released Wednesday.

They are Navy Capt. Kenneth L. Coskey and Navy Cmdr. Edwin Arthur (Ned) Shuman III of Virginia Beach, Air Force Col. Thomas H. Kirk Jr. of Portsmouth, Air Force Col. Kenneth R. Fleenor of Hampton, Air Force Lt. Col. Robert R. Craner of Hampton and Cohoes, N.Y., and Navy Lt. Cmdr. Dale W. Doss.

Doss' wife and children lived in Virginia Beach at the time of his capture in 1968, but they have since moved to Tennessee.

Also on the list is Navy Lt. Cmdr. John S. McCain III, son of retired Adm. John S. McCain Jr., former commander of U.S. forces in the Pacific.

The elder McCain served a number of years in Tidewater during his naval career. His aviator son, who was shot down over North Vietnam Oct. 26, 1967, lists his home as Jacksonville, Fla.

After their release Wednesday, the men will undergo preliminary medical checks and will be fitted for uniforms at Clark Air Base in the Philippines. Then they will be flown to military hospitals near their homes for extended physical and intelligence evaluations.

One goal of the intelligence checks is to gather information about Americans missing in Vietnam.

It is expected that the Navy men from Tidewater will receive their extended treatment at Portsmouth Naval Hospital, where eight former POWs are being checked.

The Air Force men probably will be taken to another hospital near Washington. However, Kirk's father, retired Norfolk Naval Shipyard progressman Thomas H. Kirk Sr., is hoping that spe-

(See Tidewater, Page A7)

Army CWO Joseph Rose III and his bride, Donna, pose for photographers after their wedding Monday in the Valley Forge General Hospital chapel in Phoenixville, Pa. Rose, a former POW, has been a patient at the facility since his release in Vietnam last week. The newlyweds are from Morgantown, W. Va. Story is on Page A7. (AP)

Downey is accompanied by U.S. officials at Hong Kong. (AP)

A 20-Year Prisoner Sees Ailing Mother

NEW BRITAIN, Conn. (AP) — CIA agent John T. Downey, the last known POW of the Korean War, reached the bedside of his critically ill mother Monday night after a flight halfway around the world.

Downey, released Sunday by the Chinese who had imprisoned him for espionage for more than 20 years, arrived in his native Connecticut at 8:50 p.m. He was rushed by motorcade to New Britain General Hospital, where his mother, Mary V. Downey, 75, has been under treatment since last Wednesday after suffering a stroke.

Gov. Thomas J. Meskill led a contingent of dignitaries that greeted Downey at the airport. Meskill and Downey embraced as Downey emerged from the plane in the glare of television lights.

Downey moved quickly to a car, but as newsmen crowded around, the vehicle was stopped and Downey emerged with his brother, William, who had flown to Clark Air Base in the Philippines to accompany Downey on the flight home. A woman broke from the crowd and rushed toward Downey, but was turned back by state police.

"Thank you very much," Downey said. "It's great to be back. I just want to thank everyone for being so kind."

As his motorcade pulled up outside the hospital, a crowd of about 300 persons surged around the cars, cheering. Two young people held aloft a sign that read: "Welcome Home John Downey."

He alighted from the car and was quickly surrounded by newsmen. Asked what his first thought was when he stepped onto Western soil, he replied: "I was home."

Downey was escorted quickly into the hospital lobby where he stopped momentarily to shake hands with several nurses. Then he entered an elevator for the brief ride to the fourth floor to be with his mother.

Meskill, a long-time friend of the Downey family, said before any consideration is given to celebrating Downey's return, Connecticut should "let him adjust to America 22 years later."

Concerning a celebration, Meskill said, "Whatever we can think of is not enough for what he has done for us."

Noting that he had seen Downey on television as he left China, the governor added, "I think we've aged more than he has."

Today's Chuckle

Honest, there's a sign on an important state highway that says: "When this sign is under water, road is impassable."

No Testimony From Nixon's Staff

Committee Requests to Be Rejected

By GAYLORD SHAW

WASHINGTON (AP) — Expanding on White House executive-privilege guidelines, President Nixon said Monday that past and present members of his personal staff will turn down requests to appear formally before congressional committees.

Access To Indians Shut Off

By JOHN A. LUNDQUIST

WOUNDED KNEE, S.D. (AP)—Federal law enforcement officers blocked Wounded Knee again Monday and vowed to cut off a pipeline of reinforcements, ammunition, and food to militant

ABOVE: Christening of USS Eisenhower (CVN 69), second of the Nimitz-class carriers, October 11, 1975. Courtesy Virginian-Pilot archives

RIGHT: As Saigon fell in April 1975, U.S. Marine helicopters evacuated hundreds of Americans and thousands of Vietnamese. Courtesy Virginian-Pilot archives

BELOW: The amphibious assault ship USS Tarawa (LHA 1) during sea trials, June 1976. In addition to carrying helicopter squadrons, Tarawa also took part in an early experiment with AV-8A Harrier jets. Courtesy Official US Navy Photo, Virginian-Pilot archives

Showers Likely
High 70, Low 55
Details, Page A3

The Virginian-Pilot

110th Vol. No. 160
Price 15 Cents

Norfolk, Portsmouth, Virginia Beach, Chesapeake, and Suffolk, Virginia, Wednesday, April 30, 1975

Hope Turns Bitter for S. Vietnamese Left Behind

By PETER ARNETT

SAIGON (AP)—Ten years ago I watched the first U.S. Marines arrive to help South Vietnam. They were greeted on the beaches by pretty Vietnamese girls in white-silken robes who draped flower leis around their necks.

The writer of this dispatch, winner of a Pulitzer Prize for reporting from Vietnam, was there when American soldiers first came in force to Indochina. He was there Tuesday when the Americans left, one of three Associated Press correspondents who elected to remain in Saigon after the evacuation.

A decade has passed.

And on Tuesday I watched U.S. Marines shepherding the last Americans out of South Vietnam. They were the same, clean-cut-looking young men of a decade ago.

But the Vietnamese were different.

Those who didn't have a place on the last helicopters out of Saigon—and there were thousands of them left behind—hooted, booed, and scuffled with the U.S. Marines guarding the landing zones.

Some Vietnamese threw themselves over walls and wire fences, only to be thrown back by the Marines.

Bloodshed was avoided seemingly only by good luck and bad aim on the part of some angry Vietnamese soldiers who shot at a few buses and departing helicopters.

But the entire frantic dash from Saigon by the Americans—and the bitter resentment of the thousands of Vietnamese who couldn't go—seemed a sad but accurate reflection of what relations between Americans and Vietnamese had come to in the 10 years since those flowers were gladly given to the Marines.

Americans and the South Vietnamese used to get along pretty well. That was in the days when the U.S. Marines first arrived in Vietnam imbued with a determination to see the war through.

The South Vietnamese army, dispirited then, watched with wonder as first the Marines and then the paratroopers and the American infantry came to steaming-hot Vietnam to trudge the coastal plains and mountain valleys in a punishing, unfamiliar environment.

Vietnamese officers began aping the American way. The Americans seemed always to have better-pressed uniforms and more detailed maps and diagrams.

Nearly 20,000 Vietnamese officers flew to the United States for education or advanced training, and they returned with American slang expressions and an American taste for firepower and massive military supplies.

But something went wrong along the way. To win a war like Vietnam, the subject to study was not the American way but the Communist Vietnamese way.

(See U.S. Exit, Page A4)

Saigon Regime Surrenders

Ford Says U.S. Ends War Era

By BARRY SCHWEID

WASHINGTON (AP)—The United States completed the month-long evacuation of more than 6,000 Americans and about 56,000 South Vietnamese from Saigon on Tuesday as President Ford called on the nation "to close ranks, to avoid recrimination about the past."

When word was flashed to the White House that Ambassador Graham Martin and the last evacuees had been airlifted from South Vietnam, Ford declared: "This action closes a chapter in the American experience."

About 6,500 were rescued on the last day, about 1,000 of them Americans. Ford said it was now time "to look ahead to the many goals we share and to work together on the great tasks that remain to be accomplished."

Delays blamed on bad weather, pilot fatigue, and difficult helicopter landings stretched out the day's withdrawal, which marked the end of U.S. involvement in the Vietnam War—a war that Secretary of State Henry A. Kissinger acknowledged did not meet U.S. objectives.

But, sharing the President's sentiments, Kissinger said, "It is a time to heal wounds, to look to our international obligations and to remember that peace and progress in the world has depended importantly on American commitment and American conviction. . . ."

Two U.S. marine pilots were lost in the final effort when their helicopter fell into the South China Sea.

White House press secretary Ron Nessen said the operation was extended several hours because "a lot more Vietnamese were taken out than had been planned."

He said other reasons were occasional bad weather and pilot weariness. In addition, only two helicopters could go in at one time to make pickups from the embassy roof and the parking lot, he said.

In a briefing, Secretary of State Henry

(See Chapter, Page A4)

A U.S. Marine, wearing helmet and flak jacket, tries to comfort a boy in Saigon. (UPI)

Helicopter Armada Lifts Last Evacuees

By GEORGE ESPER

SAIGON (AP)—South Vietnam declared unconditional surrender to the Vietcong early today, ending 30 years of warfare.

President Duong Van (Big) Minh spoke to the nation just hours after an armada of U.S. Marine helicopters had completed an emergency evacuation of nearly 900 Americans and thousands of Vietnamese from the besieged capital.

Minh, a retired general and neutralist, was named president Monday in a desperate and unsuccessful attempt to negotiate a peace with the Communist leaders.

In a five-minute radio address, Minh said "The Republic of Vietnam policy is the policy of peace and reconciliation, aimed at saving the blood of our people. I ask all servicemen to stop firing and stay where you are. I also demand that soldiers of the Provisional Revolutionary Government (Vietcong) stop firing and stay in place.

"We are here waiting for the Provisional Revolutionary Government, to hand over authority in order to stop useless bloodshed."

Gen. Nguyen Huu Hanh, deputy chief of staff, then went on the air to order all South Vietnamese troops to carry out Minh's orders. "All commanders must be ready to enter into relations with commanders of the Provisional Revolutionary Government to carry out the cease-fire without bloodshed," he said.

There was no immediate reponse by the Vietcong.

As they spoke, Saigon fell silent and shellfire subsided along the northern rim where Vietcong gunners had been bombarding the airport.

Saigon police and militiamen remained at their posts indicating the Communist-led troops had not yet entered the city.

Some South Vietnamese officers complained that the evacuation of Americans had caused panic in the military, with many top army officers and most of the air force fleeing.

But it had been obvious that the capital would fall. More than a dozen North Vietnamese-Vietcong divisions were ringing Saigon, which was defended by less than one division of demoralized green troops.

Associated Press special correspondent Peter Arnett, touring the city, reported that nervous soldiers fired occasionally into the air but he saw no dead or wounded. Soldiers near the radio station at the northeastern edge of town said Communist-led forces had moved up to the Saigon River bridge and were poised to enter the city.

Streets around the abandoned U.S Embassy and ambassador's residence were littered with papers and broken furniture left behind by looters who charged in after the Americans left. The building was then burned.

Americans going to assembly points for the emergency evacuation dodged random shots fired by bitter South Vietnamese soldiers and fought off desperate civilians trying to go with them.

The massive shelling of Tan Son Nhut air base and an order by Minh for the American defense attache and his staff to get out led to President Ford's order to remove all Americans, with only a handful of newsmen and relief workers remaining voluntarily.

America's 30-year involvement in Vietnam ended in tears, panic, looting, and rage.

Embittered Saigon soldiers, feeling betrayed by America's withdrawal, fired shots at buses carrying evacuees to the embassy and at the helicopters skimming overhead.

Thousands of civilians, fearing they

(See Americans, Page A4)

Not in Vain, Schlesinger Says to Men

WASHINGTON (UPI)—As the last Americans were being evacuated from Vietnam Tuesday, Defense Secretary James R. Schlesinger said the long U.S. war effort had "failed," but he added that the attempt had not been without purpose.

In a message to 2.1 million members of the armed forces, Schlesinger said: "It will be stated that the war itself was futile. In some sense, such may be said of any national effort that fails.

"Yet our involvement was not purposeless. It was intended to assist a small nation to preserve its independence in the face of external attack and to provide at least a reasonable chance to survive.

"That Vietnam succumbed to powerful external forces vitiates neither the explicit purpose behind our involvement nor the impulse of generosity toward those under attack that has long infused American policy."

"Under circumstances more difficult than ever before faced by our military services, you accomplished the mission assigned to you by higher authority," he said. "I salute you for it."

Deliveries Delayed

Delivery of Tuesday's Virginian-Pilot was delayed in some areas because of electrical problems in our mechanical departments.

Some Virginian-Pilot carriers received their papers as much as three hours late.

In many instances, they were received too late for the carriers to deliver before school time. Those papers were delivered to subscribers after the carriers returned from school.

The Virginian-Pilot regrets the delay of service to our readers.

The electrical problems have been fixed and normal delivery schedules have been resumed.

To Come to Norfolk; AF Base to Be Shut

Greeks to Close Ports to U.S. Ships

From Virginian-Pilot Wire Reports

The Greek home-porting of five American warships, the brainchild of the former chief of naval operations, Elmo R. Zumwalt, will be terminated by an agreement between the United States and Greece, it was announced Thursday.

The Navy said all five ships will be brought to Norfolk.

The agreement, announced by the United States and Greece, also calls for the closure of a U.S. Air Force base in Greece. In addition, American military facilities remaining in Greece will be placed under Greek commanders.

The changes are the result of Greek-American negotiations that began Feb. 15. No date for implementing the moves has been announced.

The issue of U.S. bases in Greece arose after Greece announced its intention to quit the military sector of the North Atlantic Treaty Organization (NATO). Several U.S. installations were commanded jointly by and NATO.

The five warships have been in Greece under an arrangement concluded two years ago with the former Greek military regime. They are the guided-missile destroyer Sampson, destroyer escort Vreeland, and destroyers W. M. Wood, Manley, and Barry.

The vessels, as well as an estimated 3,200 American naval personnel, have been stationed near Piraeus.

Adm. Zumwalt had said that basing U.S. ships overseas would improve the morale of sailors by allowing them to be near their families.

Before the Greek home-porting arrangement, the warships went periodically from East Coast bases to the Mediterranean. The families of crewmen remained behind. However, changing the home port of the

(See American, Page A4)

In Today's Pilot

Today's Chuckle

Let's face it—we'll never really crack down on air pollution until it interferes with our television reception.

LEFT: Grand opening of the Oceana Naval Air Station commissary, 1978. Courtesy Official US Navy Photo, NAS Oceana archives

BOTTOM LEFT: Capt. James H. Flatley, commanding officer of the USS Saratoga (CV 60), cuts a cake in celebration of his 1,500th carrier landing. He achieved this mark in a F-4 Phantom in 1980. His son, Midshipman James H. "Seamus" Flatley IV (far left), was in the back seat. He, too, became a Navy pilot. Courtesy The Amy Waters Yarsinske Collection

BELOW: Capt. and Mrs. William A. Rockwell slice the cake at a change of command ceremony at Norfolk Naval Air Station, 1980. Looking on are Navy Assistant Secretary Joseph A. Doyle and Capt. Roderick P. Crawford. Courtesy Mary Lou Olson (Rockwell)

VA-35

ABOVE: Crowds gather for the homecoming of USS Nimitz, May 26, 1980. A month earlier, eight RH-53D Sea Stallion helicopters took off from the carrier in an attempt to rescue American hostages in Iran. Bad weather, malfunctions and a collision ended the mission in failure. Courtesy Virginian-Pilot archives

ABOVE: Crews employ a crane during construction of the aircraft carrier Theodore Roosevelt (CVN 71) in Newport News, 1981. Courtesy Virginian-Pilot archives

LEFT: Stephen P. Davis, a photographer's mate, takes a self portrait as he flies in the back seat of a TA-4J Skyhawk attack trainer aircraft, November 1981. Courtesy Lt. Cmdr. Stephen P. Davis, (Ret.)

OPPOSITE: Members of Attack Squadron 35 return home a day ahead of the Nimitz. Courtesy Virginian-Pilot archives

ABOVE: Sailors from the USS Independence, file through the crowd to meet up with family and friends at a Norfolk homecoming, April 11, 1984. Courtesy Virginian-Pilot archives

TOP LEFT: Mothers with babies born during deployment wait at the homecoming of the USS America in Norfolk, June 2, 1983. Courtesy Virginian-Pilot archives

LEFT: Lt. Cmdr. Chuck Scott jumps a rope barrier to get to his 6-year-old daughter, Laura, May 1984. Courtesy Virginian-Pilot archives

GOOD MORNING

The Virginian-Pilot AND THE LEDGER-STAR

SUNDAY MAY 18, 1986

Complete index/A2

Mission accomplished: Crew jubilant after Libyan attack

Quiet life hides link to drug ring

Neighbors shocked by dentist's arrest

A day for warm memories

Assad vows he won't yield to U.S. threats

Syrian leader condemns raid against Libya

"Although we wish all kinds of disasters to befall Israel, ... we refuse to carry out such acts against civil aviation. We condemn the hijacking or exploding of civilian aircraft."

Hafez Assad

The Gramm-Rudman ax is becoming a boomerang

ANALYSIS

Pearisburg mayor withstands test of time

Local News

CLASSIFIED/D5

Maury: Alumni remember the early years

Fliers welcomed home from Libyan mission

Go for leadership, grads urged

Penn Foundation executive speaks at Norfolk State

Social workers suffer with children saved from abuse

ABOVE: The Norfolk-based USS Saipan (LHA 2) during Northern Wedding, a NATO military exercise, in 1986. The second of the Tarawa class of amphibious assault ships, the Saipan could launch and direct a two-pronged attack, by helicopter from its flight deck and by landing craft from the sheltered well deck in the stern. Courtesy Official US Navy Photo, Virginian-Pilot archives

TOP LEFT: A sailor looks out from a hole in the aircraft carrier USS Coral Sea after it collided with an Ecuadoran tanker off the coast of Cuba in April 1985. Courtesy Virginian-Pilot archives

LEFT & FAR LEFT: In May 1986, the carrier Coral Sea headed home after a deployment that included a predawn strike on Libya by 11 A6-E Intruder bombers. F/A-18 Hornets, early warning radar planes and KA-6 aerial tankers also participated. Courtesy Virginian-Pilot archives

ABOVE: USS Nimitz (CVN 68) pier side with Carrier Air Wing 8 prior to getting underway to the West Coast, 1987. Courtesy Jeffrey Loncto

TOP RIGHT: Four CH-46 Sea Knight helicopters from Helicopter Combat Support Squadron 8, the Dragon Whales, fly in formation over the destroyer USS Arthur W. Radford at Norfolk Naval Station, 1988. Courtesy Lt. Cmdr. Stephen P. Davis, (Ret.)

RIGHT: USS Theodore Roosevelt, the Navy's newest aircraft carrier at the time, heads back to Norfolk, after successful completion of shock trials, 1987. Courtesy Official US Navy Photo, Virginian-Pilot archives

LEFT: An officer stares through a hole in the aircraft carrier Dwight D. Eisenhower following a collision with an anchored Spanish collier in Norfolk, 1988.
Courtesy Virginian-Pilot archives

BELOW: A-6 Intruder pilots arrive at Oceana Naval Air Station after a Mediterranean deployment aboard the USS Coral Sea, March 1988.
Courtesy Official US Navy Photo, Virginian-Pilot archives

ABOVE: A C-2A Greyhound cargo aircraft from Fleet Logistics Support Squadron 40, the Rawhides, flies over the carrier piers at Norfolk Naval Station, 1989.
Courtesy Lt. Cmdr. Stephen P. Davis, (Ret.)

ABOVE: "Today" host Bryant Gumble takes a strategic spot near a heater on the deck of the USS Theodore Roosevelt in Norfolk to explain the stateside effect of the crisis brewing in the Persian Gulf, November 1990. About a month later the carrier deployed, arriving just in time for the beginning of air strikes on Iraq. Courtesy Virginian-Pilot archives

BELOW: Airman D. Allen and Petty Officer 3rd Class Jeff Davis add a "kill mark" to an aircraft after a combat mission in Iraq, 1991. They're with Fighter Squadron 14, the Tophatters, aboard the USS John F. Kennedy. Courtesy Jeffery Davis

ABOVE: The christening of the USS George Washington (CVN 73), July 21, 1990. With a flight deck area of 4.5 acres, the ship is equipped with four catapults, four aircraft elevators and 900 miles of wiring and cables. Courtesy Virginian-Pilot archives

BELOW: Carrier Air Wing One returns to Oceana Naval Air Station from the USS America following Operation Desert Storm, 1991. Courtesy Charlie Meads/The Virginian-Pilot

RIGHT: An aviator returns home to applause at Oceana Naval Air Station with other members of Carrier Air Wings 3 and 17 after serving in Operation Desert Storm, March 1991.
Courtesy Official US Navy Photo, NAS Oceana archives

BOTTOM LEFT: Lt. Ronald Lee "Rhino" Wise, a pilot with Attack Squadron 75, the Sunday Punchers, reunites with his son, Alex, after the squadron's fly-in on March 27, 1991. Roughly six years later, Wise made the Navy's final flight in an A-6 Intruder, then transitioned to the F-14 Tomcat. He was killed in July 1998 in a collision with another Tomcat over the Mediterranean Sea.
Courtesy Official US Navy Photo, Amy Waters Yarsinske Collection

BOTTOM RIGHT: Lieutenants Robert Wetzel (left), Lawrence R. Slade (center), and Jeffrey N. Zaun, attend a ceremony in their honor at Oceana. The aviators were shot down and taken prisoner by the Iraqis during Operation Desert Storm. They were three of 10 allied prisoners-of-war repatriated through the Red Cross on March 4, 1991, a few days after the cease-fire was declared.
Courtesy Official US Navy Photo, Amy Waters Yarsinske Collection

ABOVE: An F-14 Tomcat from Fighter Squadron 41, the Black Aces, launches from the USS Theodore Roosevelt (CVN 71) on a reconnaissance escort mission for Operation Provide Comfort, protecting Kurds in northern Iraq, 1991. Courtesy Virginian-Pilot archives

TOP LEFT: USS Theodore Roosevelt at sea during Operation Provide Comfort, 1991.
Courtesy Official US Navy Photo, Virginian-Pilot archives

LEFT: An S-3B Viking surveillance and targeting aircraft launches from the Theodore Roosevelt during a direct support mission over northwest Iraq, 1991.
Courtesy Virginian-Pilot archives

'THE LIBERATION OF KUWAIT HAS BEGUN'

INSIDE: Special Coverage

The Virginian-Pilot

THURSDAY JANUARY 17, 1991 — Serving southeastern Virginia and northeastern North Carolina — 126th Year 1991, No. 38 — 50 CENTS

Allies rain fury on Iraq

2nd wave of strikes pounds military targets

DESERT STORM STRIKES BAGHDAD

No reports of U.S. jets being hit

Both fear, relief greet news of raid

Military beefs up security at installations, hospital

This time, we've got eerie ringside seats

GUY FRIDDELL

SPECIAL COVERAGE OF WAR IN THE GULF

The Virginian-Pilot

MONDAY JANUARY 21, 1991 — Serving southeastern Virginia and northeastern North Carolina — 126th Year, No. 40 — 50 CENTS

Iraq displays 7 captured allied pilots on TV

Navy aviator from Beach among POWs

Lt. Zaun is not a traitor, ex-captives say

Lt. Jeffrey N. Zaun — Asked family to pray for peace

WAR UPDATES

Patriots destroy 9 Scuds

PATRIOT MISSILE: Antidote for Scuds

Parts of the Patriot missile

RANGE OF IRAQI SCUDS

Saudis get a scare, but no injuries

Air raids isolate 695,000 Iraqi troops

Experts unsure if attacks inflicting many casualties

LEFT: As Operation Desert Storm began on January 16, 1991, military installations in Hampton Roads heightened security. Courtesy Virginian-Pilot archives

BELOW: Four days after the air strikes began, Iraq aired interviews with seven captured American pilots, including Navy Lt. Jeffrey N. Zaun of Virginia Beach. Courtesy Virginian-Pilot archives

ABOVE: The aircraft carriers USS George Washington (CVN 73), USS Theodore Roosevelt (CVN 71) and USS Dwight D. Eisenhower (CVN 69) rendezvous for a photo off the coast of Virginia, 1992. Courtesy Official US Navy Photo, Virginian-Pilot archives

RIGHT: On January 21, 1991, two Air Force A-10 Thunderbolts guided a helicopter to rescue a downed Navy pilot inside Iraq. Courtesy Virginian-Pilot archives

BELOW: The aircraft carrier USS John F. Kennedy (CVA 67), 1990s. Courtesy Official US Navy Photo, Virginian-Pilot archives

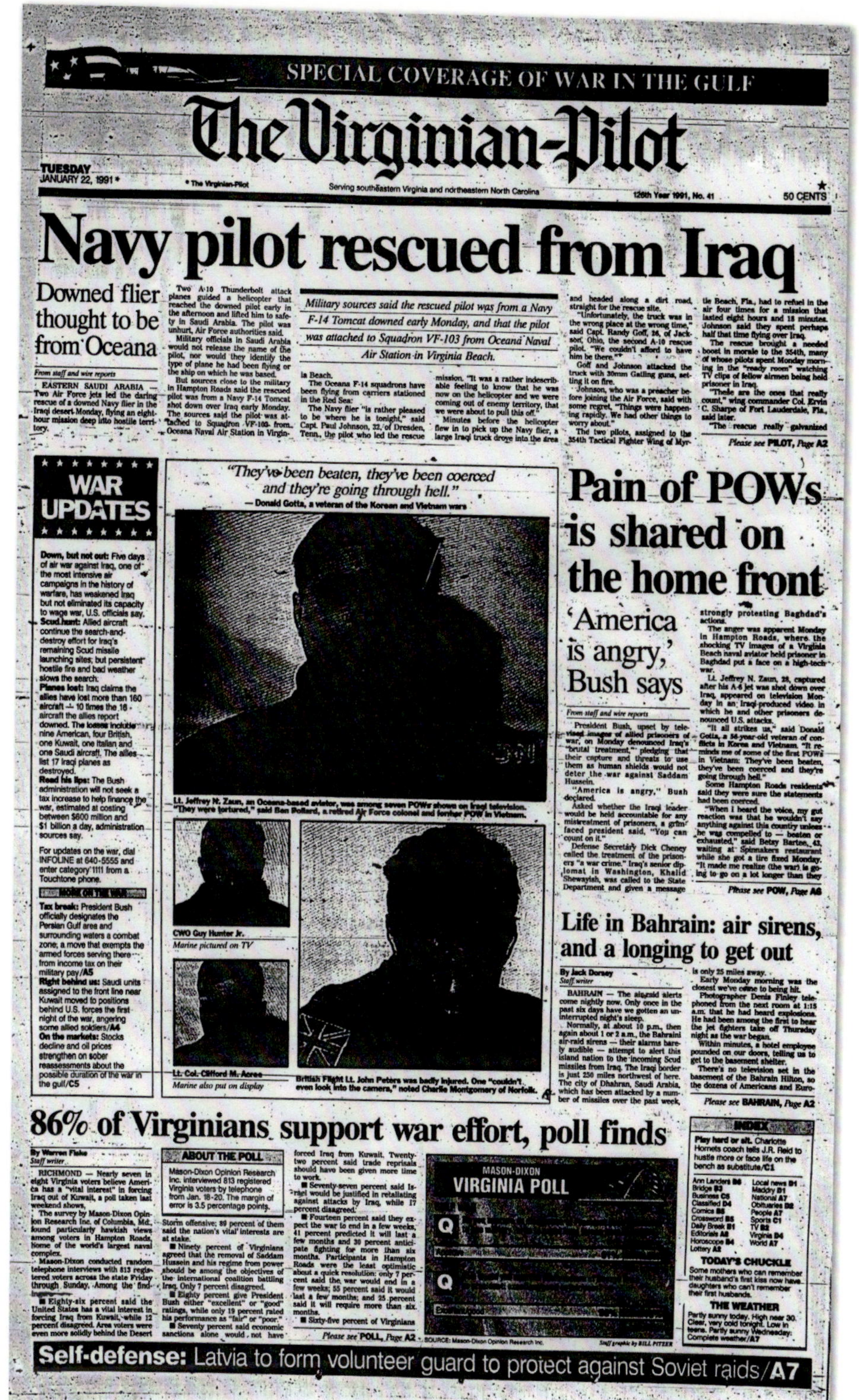

SPECIAL COVERAGE OF WAR IN THE GULF

The Virginian-Pilot

TUESDAY, JANUARY 22, 1991 • The Virginian-Pilot • Serving southeastern Virginia and northeastern North Carolina • 126th Year 1991, No. 41 • 50 CENTS

Navy pilot rescued from Iraq

Downed flier thought to be from Oceana

From staff and wire reports

EASTERN SAUDI ARABIA — Two Air Force jets led the daring rescue of a downed Navy flier in the Iraqi desert Monday, flying an eight-hour mission deep into hostile territory.

Two A-10 Thunderbolt attack planes guided a helicopter that reached the downed pilot early in the afternoon and lifted him to safety in Saudi Arabia. The pilot was unhurt, Air Force authorities said.

Military officials in Saudi Arabia would not release the name of the pilot, nor would they identify the type of plane he had been flying or the ship on which he was based.

But sources close to the military in Hampton Roads said the rescued pilot was from a Navy F-14 Tomcat shot down over Iraq early Monday. The sources said the pilot was attached to Squadron VF-103 from Oceana Naval Air Station in Virginia Beach.

Military sources said the rescued pilot was from a Navy F-14 Tomcat downed early Monday, and that the pilot was attached to Squadron VF-103 from Oceana Naval Air Station in Virginia Beach.

The Oceana F-14 squadrons have been flying from carriers stationed in the Red Sea.

The Navy flier "is rather pleased to be where he is tonight," said Capt. Paul Johnson, 32, of Dresden, Tenn., the pilot who led the rescue mission. "It was a rather indescribable feeling to know that he was now on the helicopter and we were coming out of enemy territory, that we were about to pull this off."

Minutes before the helicopter flew in to pick up the Navy flier, a large Iraqi truck drove into the area and headed along a dirt road, straight for the rescue site.

"Unfortunately, the truck was in the wrong place at the wrong time," said Capt. Randy Goff, 26, of Jackson, Ohio, the second A-10 rescue pilot. "We couldn't afford to have him be there."

Goff and Johnson attacked the truck with 30mm Gatling guns, setting it on fire.

Johnson, who was a preacher before joining the Air Force, said with some regret, "Things were happening rapidly. We had other things to worry about."

The two pilots, assigned to the 354th Tactical Fighter Wing of Myrtle Beach, Fla., had to refuel in the air four times for a mission that lasted eight hours and 18 minutes. Johnson said they spent perhaps half that time flying over Iraq.

The rescue brought a needed boost in morale to the 354th, many of whose pilots spent Monday morning in the "ready room" watching TV clips of fellow airmen being held prisoner in Iraq.

"These are the ones that really count," wing commander Col. Ervin C. Sharpe of Fort Lauderdale, Fla., said later.

"The rescue really galvanized

Please see **PILOT, *Page* A2**

WAR UPDATES

Down, but not out: Five days of air war against Iraq, one of the most intensive air campaigns in the history of warfare, has weakened Iraq but not eliminated its capacity to wage war, U.S. officials say.

Scud hunt: Allied aircraft continue the search-and-destroy effort for Iraq's remaining Scud missile launching sites; but persistent hostile fire and bad weather slows the search.

Planes lost: Iraq claims the allies have lost more than 160 aircraft — 10 times the 16 aircraft the allies report downed. The losses include nine American, four British, one Kuwait, one Italian and one Saudi aircraft. The allies list 17 Iraqi planes as destroyed.

Read his lips: The Bush administration will not seek a tax increase to help finance the war, estimated at costing between $600 million and $1 billion a day, administration sources say.

For updates on the war, dial INFOLINE at 640-5555 and enter category 1111 from a Touchtone phone.

MORE ON THE WAR

Tax break: President Bush officially designates the Persian Gulf area and surrounding waters a combat zone, a move that exempts the armed forces serving there from income tax on their military pay/**A5**

Right behind us: Saudi units assigned to the front line near Kuwait moved to positions behind U.S. forces the first night of the war, angering some allied soldiers/**A4**

On the markets: Stocks decline and oil prices strengthen on sober reassessments about the possible duration of the war in the gulf/**C5**

"They've been beaten, they've been coerced and they're going through hell."
— Donald Gotta, a veteran of the Korean and Vietnam wars

Lt. Jeffrey N. Zaun, an Oceana-based aviator, was among seven POWs shown on Iraqi television. "They were tortured," said Ben Pollard, a retired Air Force colonel and former POW in Vietnam.

CWO Guy Hunter Jr.
Marine pictured on TV

Lt. Col. Clifford M. Acree
Marine also put on display

British Flight Lt. John Peters was badly injured. One "couldn't even look into the camera," noted Charlie Montgomery of Norfolk.

Pain of POWs is shared on the home front

'America is angry,' Bush says

From staff and wire reports

President Bush, upset by televised images of allied prisoners of war, on Monday denounced Iraq's "brutal treatment," pledging that their capture and threats to use them as human shields would not deter the war against Saddam Hussein.

"America is angry," Bush declared.

Asked whether the Iraqi leader would be held accountable for any mistreatment of prisoners, a grim-faced president said, "You can count on it."

Defense Secretary Dick Cheney called the treatment of the prisoners "a war crime." Iraq's senior diplomat in Washington, Khalid Shewayish, was called to the State Department and given a message strongly protesting Baghdad's actions.

The anger was apparent Monday in Hampton Roads, where the shocking TV images of a Virginia Beach naval aviator held prisoner in Baghdad put a face on a high-tech war.

Lt. Jeffrey N. Zaun, 28, captured after his A-6 jet was shot down over Iraq, appeared on television Monday in an Iraqi-produced video in which he and other prisoners denounced U.S. attacks.

"It all strikes us," said Donald Gotta, a 54-year-old veteran of conflicts in Korea and Vietnam. "It reminds me of some of the first POWs in Vietnam: They've been beaten, they've been coerced and they're going through hell."

Some Hampton Roads residents said they were sure the statements had been coerced.

"When I heard the voice, my gut reaction was that he wouldn't say anything against this country unless he was compelled to — beaten or exhausted," said Betsy Bartee, 43, waiting at Spinnakers restaurant while she got a tire fixed Monday. "It made me realize (the war) is going to go on a lot longer than they

Please see **POW, *Page* A6**

Life in Bahrain: air sirens, and a longing to get out

By Jack Dorsey
Staff writer

BAHRAIN — The air-raid alerts come nightly now. Only once in the past six days have we gotten an uninterrupted night's sleep.

Normally, at about 10 p.m., then again about 1 or 2 a.m., the Bahraini air-raid sirens — their alarms barely audible — attempt to alert this island nation to the incoming Scud missiles from Iraq. The Iraqi border is just 250 miles northwest of here. The city of Dhahran, Saudi Arabia, which has been attacked by a number of missiles over the past week, is only 25 miles away.

Early Monday morning was the closest we've come to being hit.

Photographer Denis Finley telephoned from the next room at 1:15 a.m. that he had heard explosions. He had been among the first to hear the jet fighters take off Thursday night as the war began.

Within minutes, a hotel employee pounded on our doors, telling us to get to the basement shelter.

There's no television set in the basement of the Bahrain Hilton, so the dozens of Americans and Euro-

Please see **BAHRAIN, *Page* A2**

86% of Virginians support war effort, poll finds

By Warren Fiske
Staff writer

RICHMOND — Nearly seven in eight Virginia voters believe America has a "vital interest" in forcing Iraq out of Kuwait, a poll taken last weekend shows.

The survey by Mason-Dixon Opinion Research Inc. of Columbia, Md., found particularly hawkish views among voters in Hampton Roads, home of the world's largest naval complex.

Mason-Dixon conducted random telephone interviews with 813 registered voters across the state Friday through Sunday. Among the findings:

■ Eighty-six percent said the United States has a vital interest in forcing Iraq from Kuwait, while 12 percent disagreed. Area voters were even more solidly behind the Desert Storm offensive; 89 percent of them said the nation's vital interests are at stake.

■ Ninety percent of Virginians agreed that the removal of Saddam Hussein and his regime from power should be among the objectives of the international coalition battling Iraq. Only 7 percent disagreed.

■ Eighty percent give President Bush either "excellent" or "good" ratings, while only 19 percent rated his performance as "fair" or "poor."

■ Seventy percent said economic sanctions alone would not have forced Iraq from Kuwait. Twenty-two percent said trade reprisals should have been given more time to work.

■ Seventy-seven percent said Israel would be justified in retaliating against attacks by Iraq, while 17 percent disagreed.

■ Fourteen percent said they expect the war to end in a few weeks, 41 percent predicted it will last a few months and 30 percent anticipate fighting for more than six months. Participants in Hampton Roads were the least optimistic about a quick resolution: only 7 percent said the war would end in a few weeks; 55 percent said it would last a few months; and 25 percent said it will require more than six months.

■ Sixty-five percent of Virginians

Please see **POLL, *Page* A2**

ABOUT THE POLL

Mason-Dixon Opinion Research Inc. interviewed 813 registered Virginia voters by telephone from Jan. 18-20. The margin of error is 3.5 percentage points.

MASON-DIXON VIRGINIA POLL

SOURCE: Mason-Dixon Opinion Research Inc. *Staff graphic by BILL PITZER*

INDEX

Play hard or sit. Charlotte Hornets coach tells J.R. Reid to hustle more or face life on the bench as substitute/**C1**

TODAY'S CHUCKLE

Some mothers who can remember their husband's first kiss now have daughters who can't remember their first husbands.

THE WEATHER

Partly sunny today. High near 30. Clear, very cold tonight. Low in teens. Partly sunny Wednesday. Complete weather/**A7**

Self-defense: Latvia to form volunteer guard to protect against Soviet raids/**A7**

ABOVE: Lt. Cmdr. Brent W. Jett Jr. poses in front of an F-14 Tomcat from Fighter Squadron 74, the Be-Devilers, 1992. Jett was selected by NASA in March 1992 for astronaut training. He piloted two space shuttle missions and commanded more, aboard Endeavour and Atlantis, and logged more than 40 days in space. Courtesy Official US Navy Photo, Amy Waters Yarsinske Collection

RIGHT: Two months after their capture, three Oceana aviators taken prisoner in Iraq — Lieutenants. Lawrence R. Slade, Robert Wetzel and Jeffrey N. Zaun — returned safely to Virginia Beach. Courtesy Virginian-Pilot archives

NCAA tournament: Seton Hall, Kansas pull upsets; Arkansas, UNLV roll/**C1**

BUSINESS
Past due
Tinee Giant owes more than $1 million to its creditors/**C9**

PREVIEW
It's cool, it's 'Cats'
It's a week for big shows as a hit Broadway musical and 'Stars on Ice' both arrive

THE DAILY BREAK
Arms race
Afloat or ashore, ODU's rowing club keeps a straight course/**B1**

The Virginian-Pilot

FRIDAY
MARCH 22, 1991

Serving southeastern Virginia and northeastern North Carolina

126th Year 1991, No.84

50 CENTS

'They are American heroes'

Lts. Robert Wetzel, left, Lawrence R. Slade and Jeffrey N. Zaun are the center of attention during a ceremony Thursday welcoming them home to Oceana Naval Air Station. Each received the Purple Heart, the POW Medal and the National Defense Service Medal.

3 ex-POWs come home at last

By Kerry Dezzutti
Staff writer

BACK AT THE BEACH
Old friends. Jeff Zaun catches up on local news/A3

VIRGINIA BEACH — Most in the crowd had never met the three Oceana fliers, but they welcomed them home like old friends.

Thousands of well-wishers at Oceana wave American flags and struggle to catch a glimpse of the three returning fliers. They were eager to shake hands or simply touch the former POWs.

Please see POWs, Page A3

Gulf war vets could receive pay increase

Bill also boosts education aid

By R.A. Zaldivar
Knight-Ridder News Service

WASHINGTON — Veterans of the Persian Gulf war would receive a retroactive combat-pay increase and a more generous education allowance under a $655 million benefits package approved Thursday by Congress.

Looters, smugglers plied the Red Sea

Senior military writer Jack Dorsey is with the carrier Kennedy battle group en route home from the Persian Gulf war. He will file daily reports until the ships reach Norfolk on Thursday.

By Jack Dorsey
Staff writer

ON THE WAY HOME
Longing for land. The sailors of the cruiser Mississippi have been at sea for 18 of the past 22 months/A2

ABOARD THE CRUISER MISSISSIPPI — Time and again, in the strange and dangerous days before the shooting started, a freighter out of Jordan tried to duck past the guided-missile cruiser Mississippi.

Please see LOOTERS, Page A2

Burning tires raise pollution fears

By Jane Leonard and Linda McNatt
Staff writers

MORE ON THE FIRE
Fire watchers. People in Sussex figured the fire dump would catch fire sometime/A8

INDEX

Collision at sea. All 27 people aboard were feared dead after two Navy submarine-hunting planes collided 60 miles from San Diego/A4

Another tournament. The Norfolk State Spartanettes take their 33-0 record into a game against Southeast Missouri State tonight in the semifinals of the Division II women's basketball tournament/C1

TODAY'S CHUCKLE
They don't call it pop music because your dad likes it.

THE WEATHER
A 60 percent chance of rain today. High 70 to 75. A 50 percent chance of rain tonight. Low 50 to 55. Complete weather/A4

New Hampshire jury gets teacher's sex-murder case

By Fox Butterfield
New York Times News Service

EXETER, N.H. — In a murder case that has become New Hampshire's own real-life soap opera, the prosecution charged in closing arguments that Pamela Smart, a 23-year-old high school teacher, got her "hooks so deep into the hormones" of one of her students that he killed her husband for her.

Pamela Smart
Admitted having affair with teen

Please see TEACHER, Page A9

ABOVE: The aircraft carrier John C. Stennis (CVN 74), named for a U.S. senator, is christened in Newport News, 1993. Courtesy Virginian-Pilot archives

TOP RIGHT: Vice President Albert Gore speaks at the christening. Courtesy Virginian-Pilot archives

RIGHT: The USS John F. Kennedy (CV 67) marks its 25th anniversary in 1993 with Carrier Air Wing Three on board. Courtesy Official US Navy Photo, Amy Waters Yarsinske Collection

ABOVE: Cmdr. Jim Stark, left, smiles and Cmdr. Bob Gilman gives a thumbs up after the final flight of Attack Squadron 65, at Oceana Naval Air Station, 1993. The squadron had been active since May 1945. Courtesy Virginian-Pilot archives

OPPOSITE: F-14B Tomcats from Fighter Squadron 84, the Jolly Rogers, make a bombing run on the practice range at Vieques Island, Puerto Rico. Then the Navy's main Atlantic training site, it was shut down in 2003 after years of local protests. Courtesy Official US Navy Photo, Amy Waters Yarsinske Collection

BELOW: F-14B Tomcats from Fighter Squadron 143, the Pukin' Dogs, fly in formation over the USS George Washington (CVS 74) in the Atlantic Ocean, 1994. Courtesy Lt. Cmdr. Stephen P. Davis, (Ret.)

ABOVE: A-6E Intruders from Attack Squadron 34, the Blue Blasters, are painted to commemorate the 50th anniversary of D-Day, 1994. Courtesy Official US Navy Photo, Amy Waters Yarsinske Collection

TOP: Lt. W. Nielsen makes the first arrested landing aboard the USS George Washington in an F/A-18 Hornet strike fighter during fixed-wing flight deck certification operations off the coast of Virginia. Courtesy Official US Navy Photo, NAS Oceana archives

ABOVE: A sailor inspects the rotor head of an H-3 Sea King helicopter on the flight deck of the carrier USS George Washington, 1997. Courtesy Bill Tiernan/The Virginian-Pilot

BELOW: A plane captain hurries down the steps from the flight deck to the hangar on the USS John F. Kennedy. Some of the youngest men and women working on the flight deck are the brown shirts. They typically work battening down aircraft and are often seen carrying thick chains. Courtesy Chris Tyree/The Virginian-Pilot

ABOVE: Army Rangers exercise on the flight deck of the USS George Washington, 1994. The Rangers were on board for a joint exercise. Courtesy Virginian-Pilot archives

ABOVE: Sailors on the flight deck congratulate each other after a successful series of flight operations on the aircraft carrier USS George Washington, 1997.
Courtesy Bill Tiernan/The Virginian-Pilot

ABOVE: A woman shields her face from the sun during the christening of the carrier USS Harry S. Truman (CVN 75), 1996. Courtesy L. Todd Spencer/The Virginian-Pilot

RIGHT: The aircraft carrier USS America (CV 66) returns to Norfolk Naval Station from her last deployment, February 1996. The line handlers, who usually dressed in dungarees, wore dress blues to welcome America home. The carrier was decommissioned in August 1996. Courtesy Harry Gerwien

ABOVE: A sailor tilting seats to drain water from an early morning rain before the commissioning of the USS Harry S. Truman at Norfolk Naval Station, July 25, 1998. Courtesy Mark P. Mitchell/The Virgnian-Pilot

TOP LEFT: Sailors race to man the ship during the commissioning. Courtesy Mark P. Mitchell/The Virgnian-Pilot

LEFT: President Bill Clinton acknowledges the crowd. Defense Secretary William Cohen is at left and the Truman's commanding officer, Capt. Thomas G. Otterbein, is at right. Courtesy Mark P. Mitchell/The Virgnian-Pilot

ABOVE: Truman sailors salute at the commissioning. Courtesy Nhat Meyer/The Virginian-Pilot

ABOVE RIGHT: The USS Enterprise (CVN 65) steams at high speed while on deployment to the Mediterranean Sea, 1999. Courtesy Lt. Cmdr. Stephen P. Davis, (Ret.)

RIGHT: A stacked formation of F/A-18 Hornets, each from a different Atlantic Fleet strike fighter squadron, 1997. From front left are squadrons 37 (the Bulls), 34 (the Blue Blasters), 81 (the Sunliners), 131 (the Wildcats), 136 (the Knighthawks), 105 (the Gunslingers), 83 (the Rampagers), and 106 (the Gladiators). Courtesy Official US Navy Photo, Amy Waters Yarsinske Collection

ABOVE: James Terry, 17, with the ROTC unit at Green Run High School in Virginia Beach, watches Strike Fighter Squadron 131, the Wildcats, do a flyover at Oceana Naval Air Station, 1998. Courtesy Nhat Meyer/The Virginian-Pilot

CHAPTER FIVE

2000 to Present

THE BIGGEST ISSUES FACING NAVAL aviation in 2000 were tight maintenance budgets, a lack of replacement parts and concern about the Navy's fleet of aging planes, jets and helicopters.

But the 21st century's calm start wasn't destined to last.

After terrorists attacked the World Trade Center in New York and the Pentagon in northern Virginia, Navy aviators and crews were soon put to the test.

Within weeks of 9/11, bomb-laden F-14 Tomcats and F/A-18 Hornets were screaming off carrier decks, their ordnance headed for al-Qaeda targets in Afghanistan, while E2-C Hawkeye planes provided surveillance and battlespace management.

In 2003, the U.S. went to war with Iraq and once again, the Navy played a key role, with contributions from five carrier battle groups deployed to the Persian Gulf and eastern Mediterranean Sea. Carrier-based Tomcats, Hornets and the newest variant, the F/A-18 E/F Super Hornets, flew more than 7,000 sorties in the opening weeks of war. In addition to dropping bombs, aviators provided close-air support for soldiers and Marines on the ground.

The decade saw the retirement of three carriers: Constellation, John F. Kennedy and Kitty Hawk. And in 2006, the aviation community marked the end of another era, as the last of the iconic, swept-wing Tomcats left the service.

New airframes and updated aircraft entered the scene. In 2007, the Marines put into service the MV-22 Osprey, an aircraft with tilt-rotors that allow it to take off, land and hover like a helicopter and cruise like a plane. New sights on carrier decks include the EA-6G Growler, an updated electronic warfare plane similar to the Super Hornet, and the E2-D Hawkeye, both expected to serve for many decades to come.

OPPOSITE: Friends and family watch the Diamondbacks of Fighter Squadron 102 perform a flyover as the air wing from the USS Theodore Roosevelt returns to Oceana Naval Air Station, 2002. The following year, the squadron moved to Atsugi, Japan. Courtesy Mort Fryman/The Virginian-Pilot

ABOVE: Airman Charles Kline checks over one of Carrier Airborne Early Warning Squadron 124's E2-C Hawkeye aircraft before a month-long deployment to Puerto Rico, January 2000. Courtesy Steve Earley/The Virginian-Pilot

BELOW: The nuclear-powered aircraft carrier Ronald Reagan (CVN 76) is ready for christening on March 4, 2001. It is the ninth Nimitz-class aircraft carrier built in Newport News. Courtesy L. Todd Spencer/The Virginian-Pilot

ABOVE: Airman Jamar Hightower works on the flight deck just outside the flight deck control center of the USS Harry S. Truman on November 28, 2000, the start of the carrier's first deployment. Courtesy Steve Earley/The Virginian-Pilot

TOP: An F/A-18 Hornet is maneuvered into position for launching off the deck of the Truman, November 28, 2000. Courtesy Steve Earley/The Virginian-Pilot

LEFT: Cmdr. Bobby Rountree waves at the crowd gathered at Oceana Naval Air Station for the return of Strike Fighter Squadron 136, the Knighthawks, 2000. The squadron came home from the carrier USS Eisenhower with another Hornet squadron and two Tomcat squadrons. Courtesy Vicki Cronis/The Virginian-Pilot

FAR LEFT: Airman Hiep Nguyen, left, and Petty Officer 3rd Class Jason Edwards stand near an F-14 Tomcat on the USS Enterprise, November 8, 2001. The two sailors, who are plane captains for the aircraft, were cleaning and servicing the plane for the trip home to Virginia Beach. Courtesy Steve Earley/The Virginian-Pilot

BELOW: President George W. Bush shakes hands with sailors aboard the USS Enterprise after a speech marking the 60th anniversary of the attack on Pearl Harbor, December 7, 2001. Courtesy Genevieve Ross/The Virginian-Pilot

ABOVE: An estimated 8,000 cheering military personnel crowded the flight deck of the USS Enterprise to hear President Bush speak. Courtesy Martin Smith-Rodden/The Virginian-Pilot

ABOVE: Petty Officer 2nd Class Michael Floyd, left, a jet-engine mechanic, signals instructions for a test of the engines of an F/A-18 Hornet in the hush house at Oceana Naval Air Station. The building is designed to lessen noise aircraft maintenance.
Courtesy Bill Tiernan/The Virginian-Pilot

LEFT: Before flights begin on the aircraft carrier USS John F. Kennedy, the entire flight deck crew walks arm-to-arm across the length of the ship, looking for loose debris.
Courtesy Drew Wilson/The Virginian-Pilot

RIGHT: Aerosmith lead singer Steven Tyler tours the USS Harry S. Truman with the ship's executive officer, Ted Carter, October 2002.
Courtesy L. Todd Spencer/The Virginian-Pilot

ABOVE: Steam rises from the hot flight deck of the Truman after a cooling shower, April 2002.
Courtesy Bill Tiernan/The Virginian-Pilot

RIGHT: Petty Officer 1st Class Tony King and four teammates attempt to pull an F-14 Tomcat during the 2002 Navy Fitness Challenge aboard the Truman. Courtesy Bill Tiernan/The Virginian-Pilot

RIGHT: Airman Cory Bennett, left, and Petty Officer 1st Class Scott Sanders, both aviation ordnancemen, haul a 2,000-pound Joint Direct Attack Munition (JDAM) to an F-14 Tomcat on the USS Harry S. Truman for a bombing raid in Iraq, March 21, 2003. Courtesy Chris Tyree/The Virginian-Pilot

BOTTOM RIGHT: An F/A-18 Hornet from the Truman refuels over northern Iraq with an Air Force KC-135 Stratotanker, April 13, 2003. Courtesy Chris Tyree/The Virginian-Pilot

ABOVE: Cmdr. Marcus Hitchcock, skipper of Fighter Squadron 32, the Swordsmen, takes off from the Truman in the eastern Mediterranean Sea on March 20, 2003, the first day of the war in Iraq. In the backseat is a radar intercept officer, Capt. Mark A. Vance, commander of Carrier Air Wing Three. Courtesy Chris Tyree/The Virginian-Pilot

ABOVE: The USS Theodore Roosevelt returns from duty in Iraq, May 29, 2003. Courtesy Mort Fryman/The Virginian-Pilot

RIGHT: Dana McKenna, left, runs into the arms of her husband, Lt. Jeff McKenna of Fighter Squadron 213, the Black Lions, as he returns to Oceana Naval Air Station from Iraq, 2003. Courtesy Vasna Wilson/The Virginian-Pilot

ABOVE: Lt. Cmdr. Trey White greets his daughter, Kristin, 5, son, Michael, 3, and wife, Carla, after returning from Iraq. White flies with Strike Fighter Squadron 15, the Valions. Courtesy Vasna Wilson/The Virginian-Pilot

ABOVE: Steam shrouds the gangway as crew members of the amphibious assault ship USS Wasp (LHD 1) go aboard early on February 17, 2004, in preparation for deployment with the Wasp Expeditionary Strike Group. Courtesy Bill Tiernan/The Virginian-Pilot

BELOW: The carrier Enterprise pulls up to the pier at Norfolk Naval Station after a six-month deployment in the Persian Gulf, February 29, 2004. Courtesy Genevieve Ross/The Virginian-Pilot

ABOVE: Lt. Cmdr. Richard Kreh gets a hug from his wife Sandy as he returns to Norfolk Naval Station with the aircraft carrier USS George Washington after a six-month deployment, July 26, 2004. Courtesy Steve Earley/The Virginian-Pilot

TOP: Sailors man the rails at the Enterprise homecoming, February 29, 2004. Courtesy Genevieve Ross/The Virginian-Pilot

ABOVE: An F/A-18F Super Hornet makes a high-speed pass over Virginia Beach, breaking the sound barrier, during the Oceana Air Show, September 26, 2004. The jet is from Strike Fighter Squadron 103, the Jolly Rogers. Courtesy L. Todd Spencer/The Virginian-Pilot

ABOVE: Crowds welcome the first squadron of F/A-18 Super Hornets to arrive Oceana Naval Air Station, April 1, 2005. The Super Hornet boasts a longer range and more lethal weaponry, among other improvements. Courtesy Ecology and Environment, Inc.

BELOW: The flight line at Oceana, August 23, 2005. Courtesy L. Todd Spencer/The Virginian-Pilot

ABOVE: As the F/A-18 E and F came in, the F-14 headed for retirement. Tomcats from Fighter Squadron 11, the Red Rippers, and Strike Fighter Squadron 213, the Black Lions, make their last public flyover in Virginia Beach, September 2005. Courtesy Scott Carpentieri

LEFT: Sailors finish loading ordnance before an F-14 Tomcat's departure from the USS Theodore Roosevelt, 2006. This is the last deployment for the Tomcats. Courtesy Hyunsoo Leo Kim/The Virginian-Pilot

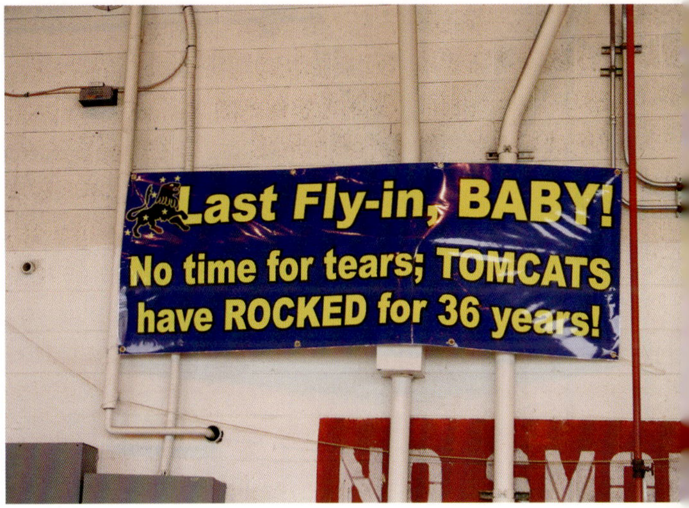

ABOVE: A sign in the hangar at Oceana Naval Air Station's final Tomcat homecoming. Courtesy Larry Wilcox

LEFT: Maneuvering an aircraft to the catapult shuttle is a precise job. Occasionally an aircraft will overshoot the shuttle. Here, flight deck personnel push an F/A-18 Hornet back for another try. Courtesy Harry Gerwien

ABOVE: Ordnance Unit personnel attach a bomb to an F-14 Tomcat before takeoff on the USS Theodore Roosevelt, 2006. Courtesy Hyunsoo Leo Kim/The Virginian-Pilot

ABOVE: A "shooter" stands near the edge of the deck as an F/A-18 Hornet is catapulted from the deck of an Aircraft Carrier participating in an exercise off the Virginia coast.
Courtesy Harry Gerwien

RIGHT: Sailors aboard the aircraft carrier USS Enterprise look for their loved ones from the flight deck of the ship as it pulls up to the pier at Norfolk Naval Station, 2004. Courtesy Genevieve Ross/The Virginian-Pilot

BOTTOM RIGHT: Cmdr. Bob Geis, skipper of Strike Fighter Squadron 211, the Checkmates, prepares for a close air support mission over Afghanistan during the first East Coast deployment of the F/A-18 Super Hornet, 2006. Courtesy Robert Geis

BELOW: A Super Hornet from the Checkmates flies above the aircraft carrier USS Enterprise, 2006. Courtesy Robert Geis

ABOVE: A Super Hornet from the Checkmates on a 2006 deployment with the Enterprise. Courtesy Robert Geis

TOP LEFT: A plane director trains a colleague as they guide a fighter jet onto the catapult aboard the aircraft carrier USS Enterprise during flight operations, 2005. Courtesy Stephen M. Katz/The Virginian-Pilot

LEFT: Lt. Bill Ellis, a launch officer or "shooter" aboard the USS Theodore Roosevelt, sports a kilt with a plaid known as "Irish revenge" to celebrate the air wing's departure as the carrier returns from an Iraq deployment. Courtesy Martin Smith-Rodden/The Virginian-Pilot

MEASUREMENTS

Length: 62 feet 9 inches

Height: 16 feet

Wingspan: 64 feet extended; 38 feet swept back

Maximum takeoff weight: 76,000 lbs.

CAPABILITIES

Top speed: 1,544 mph, more than Mach 2

Ceiling: 50,000+ feet

Range: 1,840 miles

COST

$50 million each

WHY 'TOMCAT'?

The name "Tomcat" had originally been proposed for the Grumman F7F night fighter in 1941. However, an admiral reportedly felt the name was a little suggestive and, therefore, unsuitable for a Navy fighter. The F7F became the Tigercat, on the heels of the Wildcat and the Hellcat.

When Grumman again submitted the name in 1969, the Navy said yes. It might have helped that the Navy's top three admirals were all named Tom.

BUILDER

Grumman Aircraft Corp., which was purchased by Northrop Corp. of California in 1994 and became Northrop Grumman.

"Think of driving the fastest Ferrari, the best Porsche, all in one."

– Capt. John Monroe Smith, a Tomcat pilot who retired in 1993

THAT DEPEND

After 36 years as the iconic symbol of naval av never to be seen again in the sky over Hampton R Norfolk. The twin-engine, twin-crew, swept-wing f since the retirement of the F-4 Phantom II in the c

"ROLLING TAIL"

The F-14 did not have ailerons on its moving wings. Instead, control surfaces that affect pitch and roll were located on the Tomcat's large horizontal stabilator.

ENGINES

F-14A: Two TF30-414A Pratt & Whitney afterburning turbofans, producing more than 40,000 pounds of thrust.

F-14B/D: Two F110-GE400 General Electric afterburning turbofans, producing more than 54,000 pounds of thrust.

THE SWEPT WING

Perhaps its most remarkable feature, the Tomcat's hinged wing could be controlled manually by the pilot or shifted automatically as the speed of the aircraft changed.

BLE TOMCAT

"When it launches off the deck of an aircraft carrier, it is the most feared fighter in the sky."

– www.globalsecurity.org

he F-14 Tomcat will head into the sunset Friday, aboard any of the five aircraft carriers based in as been the workhorse of carrier-based air wings ays of the Vietnam War era. A look at the Tomcat:

In its forward position, the wing allowed the Tomcat to land on an aircraft carrier at low speed. Swept at a 60-degree angle, the wing allowed the Tomcat to intercept enemy planes at high speed. The wing could be "overswept" to 75 degrees to make it easier to stuff into the hangar deck of a carrier. The wing could change its position by 7.5 degrees per second.

COCKPIT

The pilot sat in the front seat and the radar intercept officer – or RIO – sat in the rear. They shared the tasks of navigation, locating, launching weapons and electronic countermeasures.

RADAR

The large nose cone contained the AWG-9 pulse-Doppler radar system.

FUEL TANKS

The F-14 carried as many as 2,340 gallons of jet fuel in five tanks located inside the wings and fuselage.

ARMAMENT

The F-14 could track as many as 24 targets simultaneously, attack six of those targets with Phoenix missiles while continuing to scan the surrounding airspace. In all, the Tomcat carried:

ONE 20 MM VULCAN CANNON: Capable of firing 676 total rounds, mounted on the left side of the aircraft

AIR-TO-GROUND ORDNANCE IN VARIOUS CONFIGURATIONS: MK-80 series GP bombs, LGBs and JDAM

AIR-TO-AIR MISSILES: As many as six AIM-7 Sparrows, four AIM-9 Sidewinders and six AIM-54 Phoenix missiles, which could be fired at targets from a distance of about 80 miles. The Tomcat was the only U.S. plane capable of carrying the Phoenix missile.

PRODUCTION NUMBERS

Between 1970 and 1992, 633 Tomcats were built for the Navy:

493 "A" models, which initially had what was determined to be inferior Pratt & Whitney engines. Some were retrofitted with more powerful General Electric engines and converted to "B" and "D" models.

85 built as "B" models.

55 "D" models. The "B" and "D" models included the new engines and additional upgrades.

There was no "C" model – normally, the designation for a single-seat aircraft.

An additional 80 F-14A aircraft were built for Iran. Seventy-nine were delivered between 1976 and 1979, and one was retained by Grumman when the shah was overthrown in the Iranian revolution.

It took 18 to 24 months to build each Tomcat. Production of the last F-14 was in 1992.

BASES

The F-14s were split between Oceana and Miramar naval air stations, with two squadrons forward deployed to Atsugi, Japan. By 1997, however, all were based at Oceana.

RIGHT: An EA-6B Prowler radar jamming aircraft takes off from the carrier USS Harry S. Truman for a mission in northern Iraq, 2003. Courtesy Chris Tyree/The Virginian-Pilot

ABOVE: Aerobatics pilot Skip Stewart crosses the sun at the Oceana Air Show, 2007. Courtesy Larry Wilcox

LEFT: Two F8F Bearcats and a F4U Corsair pass over the Oceana Air Show, October 2010. Courtesy Warren Scott

ABOVE: An F/A-18 Super Hornet performs a tactical demonstration at the Oceana Air Show, September 20, 2008. Courtesy Bill Tiernan/The Virginian-Pilot

TOP LEFT: Evan Burgoon, 5, watches his father Lt Cmdr. Ian Burgoon of Strike Fighter Squadron 211, the Fighting Checkmates, land at Oceana Naval Air Station after a six-month deployment on the aircraft carrier USS Enterprise, December 2007. Courtesy Rich-Joseph Facun/The Virginian-Pilot

LEFT: The Red Rippers of Strike Fighter Squadron 11 make their homecoming flyover at Oceana Naval Air Station following a seven-month deployment aboard the USS Harry S. Truman, May 31, 2008. Courtesy John H. Sheally II/The Virginian-Pilot

ABOVE: Candice Knilans waits for her husband, Petty Officer 3rd Class John Knilans, to get off the USS Harry S. Truman after a seven-month deployment, June 4, 2008 in Norfolk. Courtesy Rich-Joseph Facun/The Virginian-Pilot

TOP RIGHT: Lt. Cmdr. Chad "Petey" Schumacher is welcomed home by his 3-year-old daughter, Ella, after a three-month deployment aboard the aircraft carrier USS Carl Vinson, 2009. Courtesy Mike Lynaugh

RIGHT: Lt. Mike Segrest embraces his daughters, Aubrey, 4, and Austin, 6, after returning to the United States from a deployment on the aircraft carrier USS Enterprise, July 2011. Segrest is an F/A-18 fighter pilot with the Knighthawks of the Strike Fighter Squadron 136. Courtesy Brad Vest/The Virginian-Pilot

FAR RIGHT: Petty Officer 2nd Class Michael P. Rupe kisses his girlfriend, Mary Owens, on the pier after returning from a seven-month deployment on the USS Theodore Roosevelt, April 18, 2009. Courtesy L. Todd Spencer/The Virginian-Pilot

ABOVE: Crews service helicopters aboard the USS Harry S. Truman, March 26, 2010. Courtesy Amanda Lucier/The Virginian-Pilot

LEFT: A plane captain trainee and instructor perform pre-launch checks of an E-2C Hawkeye from Carrier Airborne Early Warning Squadron 121, the Bluetails, on the USS Eisenhower, 2010. Courtesy David Mundy

ABOVE: USS Enterprise flight deck crew member Alex Lacey signals to a pilot before takeoff during a flight exercise in Atlantic Ocean, December 2010. Courtesy Hyunsoo Leo Kim/ The Virginian-Pilot

LEFT: An F/A-18 Hornet piloted by Lt. Bob Jaquith makes its final approach for landing at the Oceana Air Show, September 17, 2010. Courtesy Ken Knight

ABOVE: Lt. Raymond Guethler checks the catapult right after a takeoff from the flight deck of the USS Enterprise during exercises in Atlantic Ocean, December 2010.
Courtesy Hyunsoo Leo Kim/The Virginian-Pilot

ABOVE: Flight deck personnel on the USS Harry S. Truman manage and service aircraft, March 26, 2010. Courtesy Amanda Lucier/The Virginian-Pilot

ABOVE: An F/A-18 Super Hornet takes off from the flight deck of the USS Enterprise during exercises in the Atlantic Ocean, December 2010.
Courtesy Hyunsoo Leo Kim/The Virginian-Pilot

TOP RIGHT: Airman Alla Musa signals the flight deck crew as an F/A-18 Super Hornet is ready to take off from the flight deck of USS Enterprise during exercises in the Atlantic Ocean, December 2010.
Courtesy Hyunsoo Leo Kim/The Virginian-Pilot

BOTTOM RIGHT: A USS Enterprise flight deck crew member stands next to F/A-18 Hornet he has unchained for an exercise in the Atlantic Ocean, December 2010.
Courtesy Hyunsoo Leo Kim/The Virginian-Pilot

LEFT: Five-year-old Damien Dooley reacts with joy while watching his father, Lt. Pat Dooley, return home from seven months aboard the USS Harry S. Truman. Dooley flies with Strike Fighter Squadron 105, the Gunslingers. Courtesy Mike Lynaugh

BOTTOM LEFT: Lt. Cmdr. Chad Redmer from Strike Fighter Squadron 32, the Swordsmen, is welcomed home by his wife, Sara Stires, 2010. Courtesy Mike Lynaugh

BELOW: Lt. Cmdr. Don Calkins is welcomed home by his 3-year-old son, Peter, at Oceana Naval Air Station, December 18, 2010. Calkins flew missions in Iraq and Afghanistan with the Swordsmen. Courtesy Mike Lynaugh

AIRCRAFT ON BOARD

A Nimitz-class supercarrier can hold more than 80 aircraft on its flight deck and in its hangar. These planes are designed to handle the stress of taking off and stopping within the length of the carrier. They also must be compact enough to fit on an elevator to transport them between the flight deck and the hangar.

The F/A-18 E/F Super Hornet, a single-seat strike fighter jet, is designed to take out enemy aircraft and ground targets. Its predecessor, the F/A-18C Hornet, is also on board.

The E-2 Hawkeye is a tactical warning and control system aircraft. Its rotating radar dome updates the air wing on enemy activity.

The E/A-18G Growler, a two-seat electronic attack fighter, detects, locates and supresses enemy radar and other signals.

The C-2A Greyhound is a twin-engine cargo aircraft designed to carry mail and supplies to and from the carrier.

The EA-6B Prowler, an electronic warfare aircraft, jams enemy radar and intercepts enemy communications.

The SH-60 Seahawk, a twin-engine helicopter, attacks enemy submarines and helps in search-and-rescue operations.

GRAPHIC BY LISA MERKLIN | THE VIRGINIAN-PILOT

SOURCES: U.S. Navy, HowStuffWorks, Naval Safety Center, Federation of American Scientists, U.S. Centennial of Flight Commission

100 YEARS OF NAVAL AVIATION

THE CARRIER & ITS AIRCRAFT

A FLOATING MILITARY BASE, the aircraft carrier is the world's most powerful and flexible warship. It can supply military forces without relying on overseas bases. The carrier's design has evolved to accomodate the safe launching and landing of modern aircraft – bigger and faster than ever – at sea.

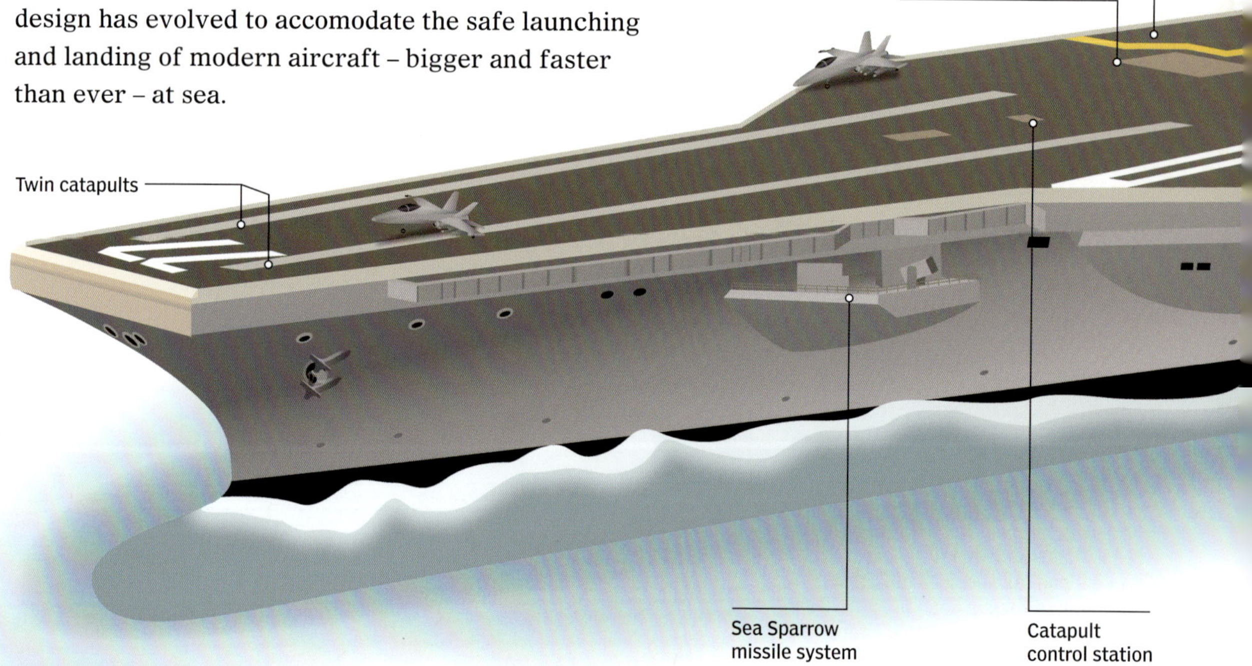

buy this poster

Purchase a color reproduction of this page at PilotOnline.com. Click on "Buy Pilot photos."

A CENTURY OF FLIGHT AT SEA

November 14, 1910 Eugene Ely takes off from an 83-foot wooden deck built on the cruiser Birmingham in Hampton Roads. Two months later, Ely lands on a specially built platform on the Pennsylvania in San Francisco.

1911 Lt. T. Gordon "Spuds" Ellyson completes his flight training, becoming Naval Aviator No. 1.

1915 Lt. Cmdr. Henry Mustin makes the first catapult launch from a ship. He flies an AB-2 flying boat off the North Carolina in Pensacola Bay, Fla.

1922 The Navy's first aircraft carrier – the Langley, a converted coal cargo ship – is commissioned in Norfolk.

1934 The Ranger, the first U.S. Navy ship to be designed and constructed as an aircraft carrier, is commissioned.

THE EARLY YEARS

1941 Carrier aircraft of the Japanese Imperial Navy launch an attack on Pearl Harbor. The three carriers of the Pacific Fleet are not present.

The first U.S. carrier offensive attacks Japanese installations in the Pacific Ocean. The Langley sinks.

1942 The U.S. carrier force defeats Japanese carriers at the Battle of Midway, the turning point of the war in the Pacific Ocean.

1944 Japanese and American naval planes clash in the Battle of the Philippine Sea – the first sea battle in which opposing ships never see each other.

1945 The carrier Franklin maneuvers closer to Japan's mainland than any other U.S. carrier. It survives an attack by a single Japanese plane.

WORLD WAR II

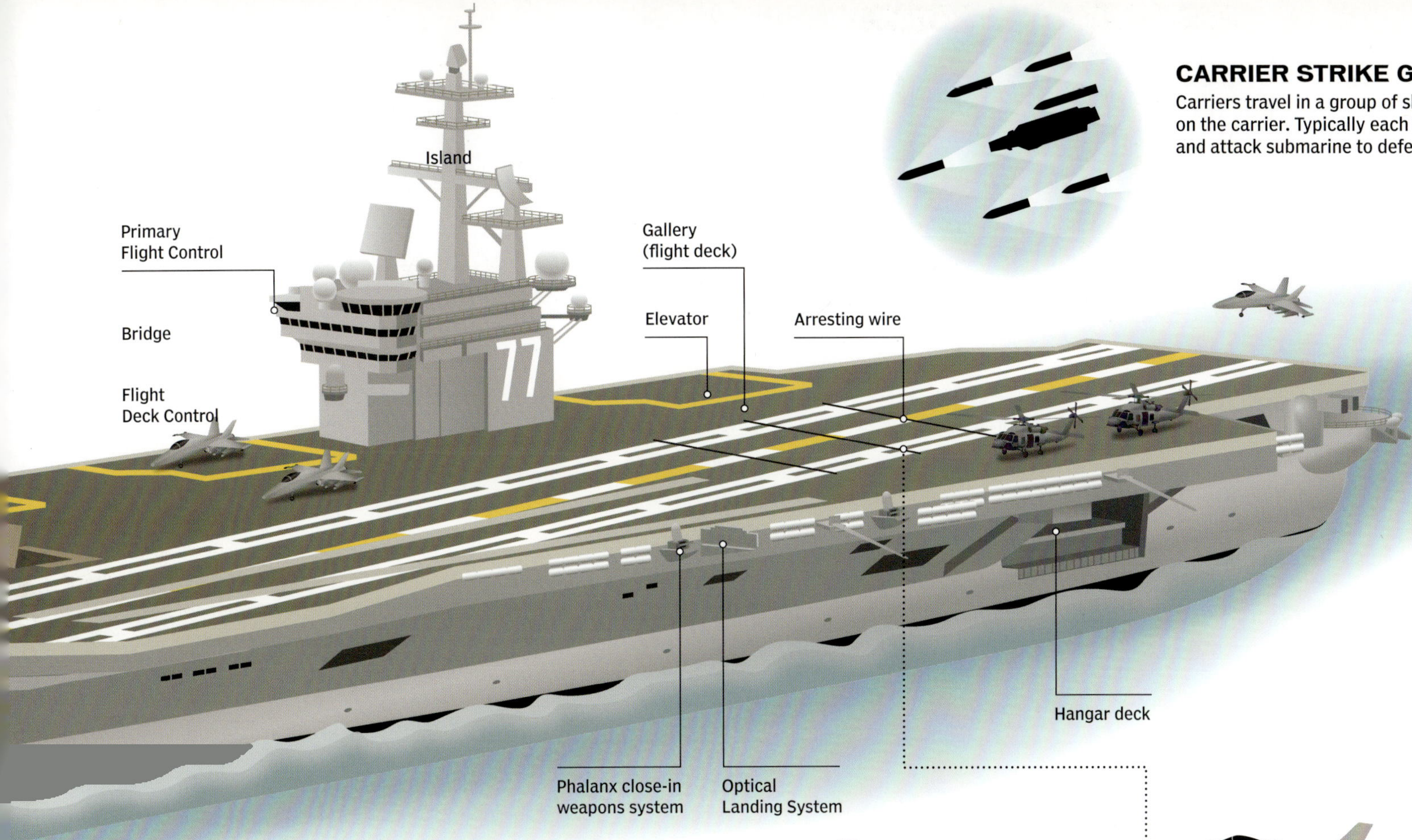

CARRIER STRIKE GROUPS

Carriers travel in a group of ships commanded by a rear admiral stationed on the carrier. Typically each has at least one frigate, destroyer, cruiser and attack submarine to defend from air, surface and submarine attacks.

THE SHIP

The newest carrier, the George H.W. Bush, below, was commissioned in 2009. The 10th and final Nimitz-class carrier, it is due to deploy this spring.

Ship personnel More than 5,000 crewmembers

Length 1,092 feet

Flight deck 4.5 acres

Weight 95,000 tons

Top speed 30 knots

Cost to build $6.2 billion

Homeport Norfolk

LAUNCHING AIRCRAFT

The catapult must bring the aircraft to a speed of 150 miles per hour within 270 feet so it can take off from the flight deck. Each of the carrier's four catapults has two steam-driven pistons that thrust the craft along the shuttle track.

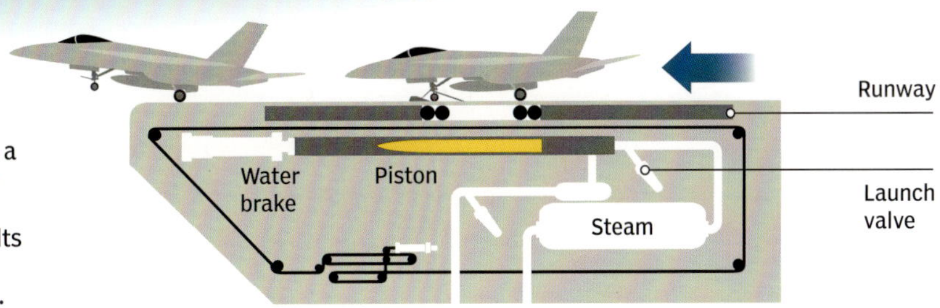

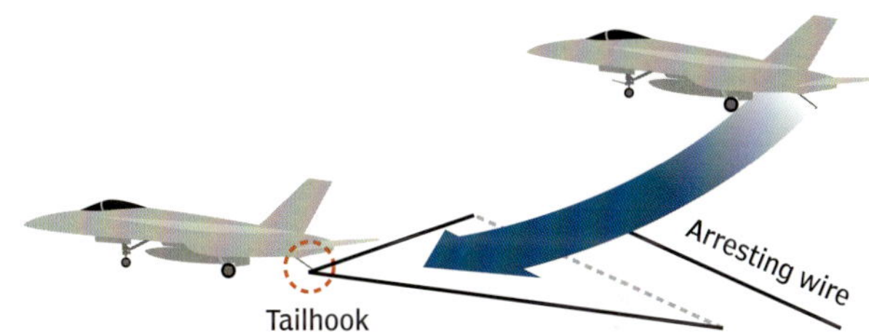

RECOVERING AIRCRAFT

Landing is one of the most difficult maneuvers a naval aviator faces. Each plane has a tailhook to grab one of the carrier's arresting wires, helping it come to a complete stop within 800 feet. The pilot guns the engine to full power in case the hook does not catch and the plane must take off again.

1947 Some World War II carriers are modernized. Improved flight decks and blast deflectors are added.

1948 Fighter Squadron 17-A becomes the first carrier-qualified jet squadron in the U.S. Navy.

President Harry S. Truman approves construction of a 65,000-ton "supercarrier" named United States. The project is canceled and the keel scrapped.

1950 With the outbreak of the Korean War, the Boxer makes record time crossing the Pacific Ocean to carry planes to combat. The trip takes 8½ days.

1952 144 aircraft from three carriers destroy the oil refinery at Aoji, North Korea, in the largest carrier raid in the Korean War.

The Antietam becomes the first carrier to be equipped with an angled deck.

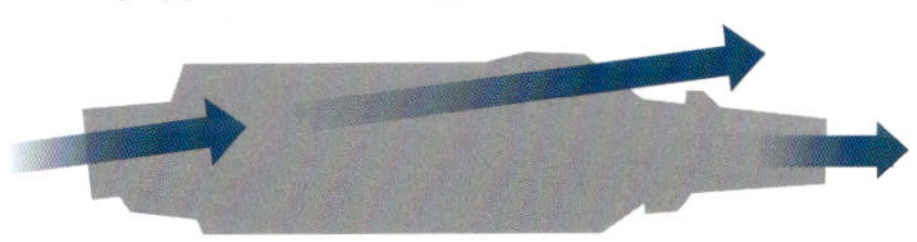

1954 The Hancock is the first carrier with steam catapults capable of launching high-performance jets.

1955 The Navy's first supercarrier, the Forrestal, is the first of four ships of its class. It suffers a huge fire in 1967.

1961 The Lake Champlain recovers Cmdr. Alan Shepard, the first American to go into space, as he completes his Freedom 7 flight.

The Enterprise becomes the first nuclear-powered aircraft carrier.

1968 The carrier Oriskany loses half its aircraft – 39 in all – in 122 days in Vietnam.

THE COLD WAR ERA

2001 Carrier-based F-14 Tomcats and F/A-18 Hornets take part in the first combat operations in Afghanistan.

TODAY
Aircraft carriers are still the fastest way to transport warplanes to crisis areas around the world. Modern supercarriers, powered by two nuclear reactors, can travel for years without refueling.

NORTHROP GRUMMAN

WHAT'S NEXT? Nimitz-class carriers are making way for the Gerald R. Ford class. The first of these new carriers, set to join the fleet in 2015, will require fewer personnel and is expected to save more than $5 billion in costs during its 50-year service life.

POST-9/11

ABOVE: An F/A-18F Super Hornet of Strike Fighter Squadron 106, the Gladiators. Courtesy Daniel Adams

BELOW: An H-3 Sea King, a twin-engine helicopter specializing in anti-submarine warfare, makes its final approach to the small flight deck of a frigate. Courtesy Harry Gerwien

ABOVE: Planes from different eras unite for a heritage flight at the Oceana Air Show, September 19, 2010. Courtesy Michelle Welliver

TOP: The Blue Angels, the Navy's elite demonstration team, fly in formation at the Oceana Air Show, September 18, 2010. Courtesy L. Todd Spencer/The Virginian-Pilot

ABOVE: Sarah Zilberman, 3, touches the image of her father, Lt. Miroslav "Steve" Zilberman, at the Naval Aviation Monument in Virginia Beach, March 2011. When mechanical problems struck his E-2C Hawkeye aircraft while returning from a mission in Afghanistan on March 31, 2010, Zilberman ordered the crew to bail out and kept the aircraft level long enough for them to do so safely. He died in the crash. Courtesy Bill Tiernan/TheVirginian-Pilot

RIGHT: The Blue Angels pass in front of the moon. Courtesy Daniel Adams

BOTTOM RIGHT: A C-2 Greyhound from Fleet Logistics Support Squadron 40, the Rawhides, sports a special paint scheme to mark the centennial of naval aviation. Courtesy Daniel Adams

BELOW: The Navy Leap Frogs parachute demonstration team perform. Courtesy Daniel Adams

APPENDIX

The Carriers

THEIR NAMES EVOKE EPIC BATTLES, presidents, and fighting ships of centuries past. Their presence allows American military might to reach across the globe. For most of the past 100 years, aircraft carriers have embodied naval aviation.

So it's only fitting to conclude this history with a salute to the carrier fleet, past and present.

Many of the ships listed here were built at the historic shipyard now known as Newport News Shipbuilding. Many were at a Norfolk pier when sailors were given the order, "Man our ship and bring her to life!"

Whether based in Norfolk, on the West Coast, or overseas, these carriers shared a mission that is unique in America's armed forces. They fought through World War II, Korea, Vietnam, the Cold War, Iraq and Afghanistan.

Some were lost in battle. Some retired to places of honor as museums or memorials. Some went to the scrap heap.

Whatever their fate, they remain in the hearts of the sailors who served aboard them and the families who waited at the pier for their return.

USS Langley (CV 1)

Commissioned: March 20, 1922
Decommissioned: February 27, 1942
Converted to AV 3; Sunk as a result of enemy action about 75 miles south of Tjilatjap, Java.

USS Lexington (CV 2)

Commissioned: December 14, 1927
Decommissioned: May 8, 1942
Sunk as a result of enemy action at the Battle of Coral Sea.

USS Saratoga (CV 3)

Commissioned: November 16, 1927
Decommissioned: July 26, 1946
Used as a test target for atomic bombs and sunk at Bikini Atoll.

USS Ranger (CV 4)

Commissioned: June 4, 1934
Decommissioned: October 18, 1946
Sold for scrapping, January 1947.

USS Yorktown (CV 5)

Commissioned: September 30, 1937
Decommissioned: June 7, 1942
Sunk due to enemy action at the Battle of Midway.

USS Enterprise (CV 6)

Commissioned: May 12, 1938
Decommissioned: February 17, 1947
Sold for scrapping, July 1958.

USS Wasp (CV 7)

Commissioned: April 25, 1940
Decommissioned: September 15, 1942
Sunk due to enemy action south of Guadalcanal.

USS Hornet (CV 8)

Commissioned: October 20, 1941
Decommissioned: October 27, 1942
Sunk due to enemy action at the Battle of the Santa Cruz Islands in the South Pacific.

USS Essex (CV 9)

Commissioned: December 31, 1942
Decommissioned: June 20, 1969
Sold for scrapping, June 1975.

USS Yorktown (CV 10)

Commissioned: April 15, 1943
Decommissioned: June 27, 1970
Established as a floating museum in Charleston, S.C., October 1975.

USS Intrepid (CV 11)

Commissioned: August 16, 1943
Decommissioned: March 15, 1974
Established as a floating museum in New York City, 1982.

USS Hornet (CV 12)

Commissioned: November 29, 1943
Decommissioned: June 26, 1970
Donated to The Aircraft Carrier Hornet Foundation for use as a museum in Alameda, Calif., May 1998.

USS Franklin (CV 13)

Commissioned: January 31, 1944
Decommissioned: February 17, 1947
Sold for scrapping, 1966.

USS Ticonderoga (CV 14)

Commissioned: May 8, 1944
Decommissioned: September 1, 1973
Sold for scrapping, 1975.

USS Randolph (CV 15)

Commissioned: October 9, 1944
Decommissioned: February 13, 1969
Sold for scrapping, 1975.

USS Lexington (CV 16)

Commissioned: February 17, 1943
Decommissioned: November 8, 1991
Now a museum in Corpus Christi, Tex.

USS Bunker Hill (CV 17)

Commissioned: May 25, 1943
Decommissioned: July 9, 1947
Sold for scrapping, 1973.

USS Wasp (CV 18)

Commissioned: November 24, 1943
Decommissioned: July 1, 1972
Sold for scrapping, 1973.

USS Hancock (CV 19)

Commissioned: April 15, 1944
Decommissioned: January 30, 1976
Sold for scrapping, 1976.

USS Bennington (CV 20)

Commissioned: August 6, 1944
Decommissioned: January 15, 1970
Sold for scrapping, 1994.

USS Boxer (CV 21)

Commissioned: April 16, 1945
Decommissioned: December 1, 1969
Sold for scrapping, 1971.

USS Independence (CVL 22)

Commissioned: January 14, 1943
Decommissioned: August 28, 1946
Sunk as target, 1951.

USS Princeton (CVL 23)

Commissioned: February 25, 1943
Decommissioned: October 24, 1944
Sunk due to enemy action in the Sibuyan Sea, Philippines.

USS Belleau Wood (CVL 24)

Commissioned: March 31, 1943
Decommissioned: January 13, 1947
Transferred to France 1953-1960. Returned and sold for scrapping.

USS Cowpens (CVL 25)

Commissioned: May 28, 1943
Decommissioned: January 13, 1947
Sold for scrapping, 1960.

USS Monterey (CVL 26)

Commissioned: June 17, 1943
Decommissioned: January 16, 1956
Sold for scrapping, 1971.

USS Langley (CVL 27)

Commissioned: August 31, 1943
Decommissioned: February 11, 1947
Transferred to France 1951-1963; Sold for scrapping, 1964.

USS Cabot (CVL 28)

Commissioned: July 24, 1943
Decommissioned: January 21, 1955
Transferred to Spain August 30, 1967; returned to private U.S. organization 1989. Sold for scrapping, 1999.

USS Bataan (CVL 29)

Commissioned: November 17, 1943
Decommissioned: April 9, 1954
Sold for scrapping, 1961.

USS San Jacinto (CVL 30)

Commissioned: December 15, 1943
Decommissioned: March 1, 1947
Sold for scrapping, 1971.

USS Bon Homme Richard (CV 31)

Commissioned: November 26, 1944
Decommissioned: July 2, 1971
Sold for scrapping, 1992.

USS Leyte (CV 32)

Commissioned: April 11, 1946
Decommissioned: May 15, 1959
Sold for scrapping, 1970.

USS Kearsarge (CV 33)

Commissioned: March 2, 1946
Decommissioned: January 15, 1970
Sold for scrapping, 1974.

USS Oriskany (CVA 34)

Commissioned: September 25, 1950
Decommissioned: September 30, 1976
Sunk off Florida coast for use as artificial reef, 2006.

Reprisal (CV 35)

Never completed.
Hull, 53% complete, was sold for scrapping, 1949.

USS Antietam (CV 36)

Commissioned: January 28, 1945
Decommissioned: May 8, 1963
Sold for scrapping, 1973.

USS Princeton (CV 37)

Commissioned: November 18, 1945
Decommissioned: January 30, 1970
Sold for scrapping, 1971.

USS Shangri-La (CV 38)

Commissioned: September 15, 1944
Decommissioned: July 30, 1971
Sold for scrapping, 1988.

USS Lake Champlain (CV 39)

Commissioned: June 3, 1945
Decommissioned: May 2, 1966
Sold for scrapping, 1972.

USS Tarawa (CV 40)

Commissioned: December 8, 1945
Decommissioned: May 13, 1960
Sold for scrapping, 1968.

USS Midway (CVB 41)

Commissioned: September 10, 1945
Decommissioned: April 11, 1992
Towed to San Diego to be used as a museum and memorial, 2004.

USS Franklin D. Roosevelt (CVB 42)

Commissioned: October 27, 1945
Decommissioned: September 30, 1977
Sold for scrapping, 1978.

USS Coral Sea (CVB 43)

Commissioned: October 1, 1947
Decommissioned: April 30, 1990
Sold for scrapping, 1993.

USS Valley Forge (CV 45)

Commissioned: November 3, 1946
Decommissioned: January 15, 1970
Sold for scrapping, 1971.

USS Philippine Sea (CV 47)

Commissioned: May 11, 1946
Decommissioned: December 28, 1958
Sold for scrapping, 1971.

USS Saipan (CVL 48)

Commissioned: July 14, 1946
Decommissioned: January 14, 1970
Sold for scrapping, 1976.

USS Wright (CVL 49)

Commissioned: February 9, 1947
Decommissioned: March 15, 1956
Sold for scrapping, 1980.

USS Forrestal (CVA 59)

Commissioned: October 1, 1955
Decommissioned: September 11, 1993
May be sunk or sold for scrapping.

USS Saratoga (CVA 60)

Commissioned: April 14, 1956
Decommissioned: August 20, 1994
Slated for scrapping.

USS Ranger (CVA 61)

Commissioned: August 10, 1957
Decommissioned: July 10, 1993
Available for donation as a museum or memorial.

USS Independence (CV 62)

Commissioned: January 10, 1959
Decommissioned: September 30, 1998
Slated for disposal.

USS Kitty Hawk (CV 63)

Commissioned: April 29, 1961
Decommissioned: May 12, 2009
In reserve.

USS Constellation (CV 64)

Commissioned: October 27, 1961
Decommissioned: August 7, 2003
Slated for disposal.

USS Enterprise (CVN 65)

Commissioned: November 25, 1961
Decommissioned: Set for fall 2012

USS America (CV 66)

Commissioned: January 23, 1965
Decommissioned: August 9, 1996
Sunk in the Atlantic Ocean off Virginia following a series of simulated attacks, 2005.

USS John F. Kennedy (CV 67)

Commissioned: September 7, 1968
Decommissioned: March 23, 2007
Available for donation as a museum.

USS Nimitz (CVN 68)

Commissioned: May 3, 1975

USS Dwight D. Eisenhower (CVN 69)

Commissioned: October 18, 1977

USS Carl Vinson (CVN 70)

Commissioned: March 13, 1982

USS Theodore Roosevelt (CVN 71)

Commissioned: October 25, 1986

USS Abraham Lincoln (CVN 72)

Commissioned: November 11, 1989

USS George Washington (CVN 73)

Commissioned: July 4, 1992

USS John C. Stennis (CVN 74)

Commissioned: December 9, 1995

USS Harry S. Truman (CVN 75)

Commissioned: July 25, 1998

USS Ronald Reagan (CVN 76)

Commissioned: July 12, 2003

USS George H.W. Bush (CVN 77)

Commissioned: January 10, 2009